Skills Coach Write Math!
How to Construct Responses to Open-Ended Math Questions
Level C

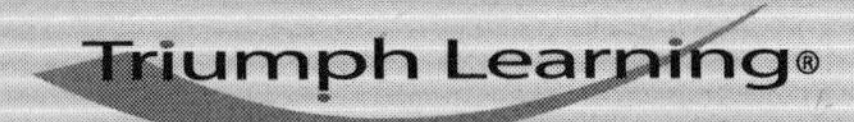

A Haights Cross Communications® Company

Skills Coach Write Math! How to Construct Responses to Open-Ended Math Questions, Level C
98NA
ISBN-10: 1-58620-909-4
ISBN-13: 978-1-58620-909-4

Author: Keith Grober
Cover Image: Myron/The Image Bank/Getty Images

Triumph Learning® 136 Madison Avenue, 7th Floor, New York, NY 10016
Kevin McAliley, President and Chief Executive Officer

Printed in the United States of America.

10 9 8 7 6

Table of Contents

Letter to the Student

Dear Student,

Welcome to the smart way to write answers to open-ended math questions. You will learn what open-ended math questions are, how to solve them, and how to score them. You will do this by reviewing modeled problems, practicing with guided questions, and answering independent problems. You will work together with your teacher, with your classmates, and with your caregiver at home.

Let's learn to **write math** the smart way!

Have fun!

1. What Is an Open-Ended Math Question?

It is a math problem with a correct answer that you can get to in different ways.

Each way is great as long as:

- it gets you to the right answer
- you show how it got you to the answer
- you explain why you chose to answer the question this way

5-step plan:

1. Read and Think • 2. Select a Strategy • 3. Solve • 4. Write/Explain • 5. Reflect

This plan will help you answer open-ended questions.

1. **Read and Think**
2. **Select a Strategy**
3. **Solve**
4. **Write/Explain**
5. **Reflect**

1. Read and Think

Read the **problem** carefully.

What **question** are you asked?

- **In your own words, tell what this problem is about.**

What are the **keywords**?

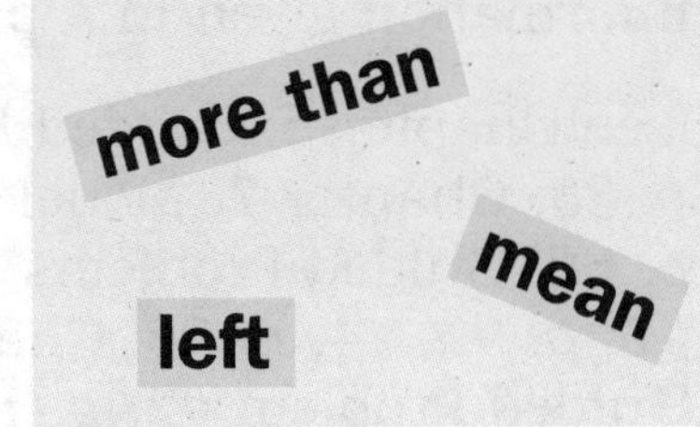

What **facts** are you given?

- **Decide what facts are needed and which ones are extra.**

2. Select a Strategy

- **How am I going to solve this problem?**
- **What strategy should I use?**

There are lots of strategies to help you solve open-ended math questions. Some are listed below and others you may come up with yourself. In parentheses, you will find where to locate examples of the strategies in use.

Draw a Picture or Graph . . .

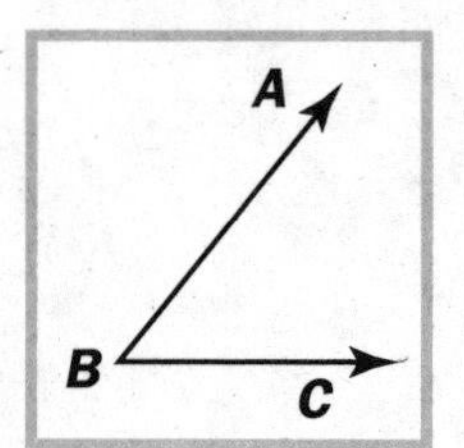

when you need to see the information given in a problem.

(see Chapter 5, Guided Problem #1, *p. 33*; Chapter 7, Modeled Problem, *p. 83*, Guided Problems #1, *p. 85*, #2, *p. 90*, #4, *p. 98*; Chapter 8, Guided Problem #2, *p. 116*)

Make a Model or Act It Out . . .

when you need to watch how the solution is found.

(see Chapter 5, Guided Problem #2, *p. 37*)

Make an Organized List or Table . . .

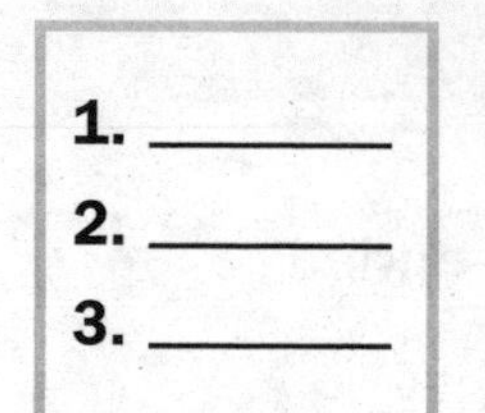

when there is a lot of information scattered throughout a problem. A list or table can help you organize your thinking.

(see Chapter 7, Guided Problem #3, *p. 94*; Chapter 9, Guided Problem #4, *p. 147*)

Look for a Pattern . . .

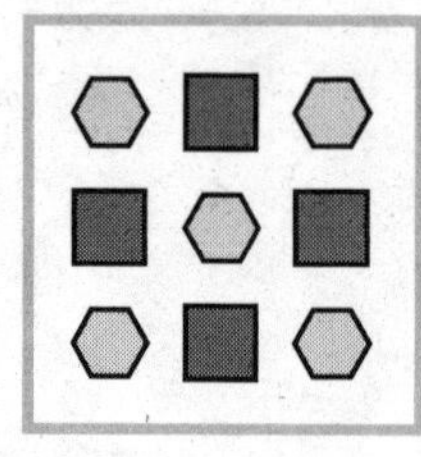

when you need to predict what comes next or find a rule. Making a list or table can often help you find a pattern.

(see Chapter 6, Guided Problem #5, *p. 76*)

Guess and Test . . .

when it is difficult to work out the answer to a problem. Make a guess. Then test it. If your guess is incorrect, use that guess to make a better guess.

(see Chapter 6, Guided Problems #2, *p. 64*, #3, *p. 68*)

Use Logical Thinking . . .

when you need to figure out how the information you have fits together.

(see Chapter 8, Modeled Problem, *p. 108*; Chapter 9, Modeled Problem, *p. 133*)

Work Backward . . .

when you know the end result or total and need to find a missing part.

(see Chapter 6, Guided Problem #4, *p. 72*)

Write a Number Sentence or Equation . . .

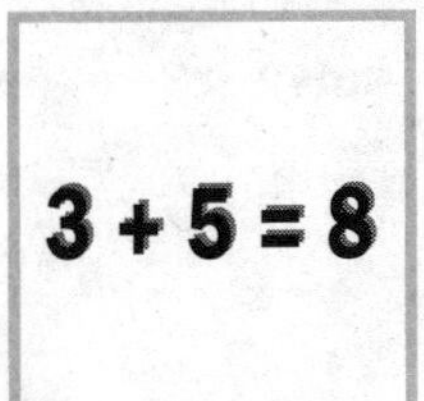

when you need to find a missing amount or show your work.

(see Chapter 6, Modeled Problem, *p. 58*, Guided Problem #1, *p. 35*; Chapter 9, Guided Problems #1, *p. 139*, #2, *p. 142*)

Divide and Conquer . . .

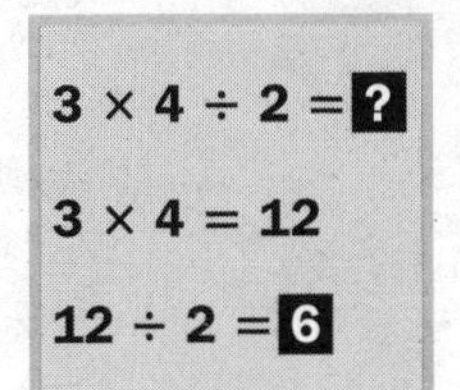

when you must solve more than one problem to find the answer to the main question. Break down the main question into steps, and solve the steps one at a time. Be sure to tell what each step is. Name the strategy you may use for each step.

(see Chapter 5, Modeled Problem, *p. 31*, Guided Problem #3, *p. 43*; Chapter 8, Guided Problems #1, *p. 110*, #3, *p. 118*; Chapter 9, Guided Problem #3, *p. 143*)

Make It Simpler . . .

20 + 40 + 50 = ?

2 + 4 + 5 = 11

Then add a zero: 110

when you must solve a complex problem with large numbers or many items.

Reduce the large numbers to small numbers, or reduce the number of items given.

(see Chapter 5, Guided Problem #5, *p. 50*)

3. Solve

After you pick your strategy, use it to solve the problem.

Use your knowledge of arithmetic and mathematics here.

- **Be very careful—you must get the correct answer. Check your arithmetic!**
- **Label your work. Use units or a sentence that explains the answer.**

4. Write/Explain

Write out an **explanation** of how you solved the problem.

Explain the strategy you chose and why you chose it.

Write your thoughts or why you solved it that way.

Don't leave out any steps.

5. Reflect

First, Review Your Work

Review what you have written.

Use the **Read It and Think** list to help you.

Read It and Think

- ☐ **Did I read the problem at least twice? Do I understand it?**
- ☐ **Did I write down the question being asked?**
- ☐ **Did I write down the keywords in the problem?**
- ☐ **Did I write down the facts that are given?**
- ☐ **Did I write down the strategy that I used?**
- ☐ **Did I solve the problem?**
- ☐ **Is my arithmetic correct?**
- ☐ **Did I explain how I solved the problem?**
- ☐ **Did I explain why I chose the strategy and how I used it?**
- ☐ **Did I include all the steps I took to solve it?**
- ☐ **Is my writing clear?**
- ☐ **Did I label my work?**
- ☐ **Does my answer make sense?**
- ☐ **Did I answer the exact question being asked?**

Then, Improve What You Wrote

How can you improve your writing?

- **Try to rewrite your answer to make it clearer, more accurate, and more complete.**

These steps might look like a lot of work. Once you start to use them, they will become very familiar and not seem so hard. You will find that these five steps work with any open-ended math question. This book will help you practice using them.

Working with this book will help you do better on math tests with open-ended questions. You will even learn how to check your answer using the **rubric** on *page 13*.

The Glossary

A Glossary is like a dictionary. It tells the meaning of words. You should learn to use the Glossary found on *pages 175–199*. It contains mathematical words you should know, including **keywords** found in the math problems throughout this book.

Sometimes a word can mean one thing in everyday life, but something else in mathematics. For example, in everyday life, the word **straight** can mean *immediately*, as in "After you wash up, go **straight** to bed." But in math, **straight** can describe a *kind of line that is not bent or curved*, as in, "The sides of a square are four straight lines."

If you are not sure what a word means mathematically in this book, look it up in the Glossary.

Tips

- If you came up with a strategy of your own that is not listed on *pages 8–9*, be sure to explain what the strategy is and why you chose it!
- Your writing must be clear. This is very important when you are taking a test. Remember, the person who reads your work must be able to figure out what you did. You can lose points if your writing is not very clear.

2. What Is a **Rubric**?

A **rubric** is a grading system used to score **open-ended** math questions. The person who scores the answers on your test uses a rubric. A rubric can also be used as a guide in answering open-ended questions. It lists things that should be found in your answer.

Throughout this book, you will use the rubric in two ways:

1. to **guide** you in answering an open-ended math question. It will remind you to write a correct, clear, complete, and thoughtful answer.
2. to score yourself as you **double-check** that your answer is complete.

2. What Is a Rubric?

Here is a typical rubric. It is used to score your work from **0** to **4**. **4** is a perfect answer.

- **You showed you knew what the problem asked.**
- **You showed you knew what facts were given, including keywords.**
- **You chose a good strategy and used it correctly.**
- **Your arithmetic or operations were done correctly.**
- **You got a correct and complete answer and labeled it.**
- **You wrote a good, clear explanation of why you chose a strategy and how you used it.**
- **You put in all the steps you used to get to your answer.**
- **You explained your thinking clearly.**

- **You showed you knew what the problem asked.**
- **You showed you knew what facts were given, including keywords.**
- **You chose a good strategy but may not have used it correctly,**
  OR
  you may have made an arithmetic error in your work.
- **You wrote an explanation of why you chose a strategy and how you used it.**
- **You might not have used all of the steps to get your answer.**
- **Your explanation was mostly clear but might not have been entirely complete.**

- **You showed you knew what the problem asked.**
- **You showed you knew what facts were given, including keywords.**
- **You chose a good strategy but may not have used it correctly,**
  OR
  you may have made an arithmetic error in your work.
- **Your answer may not be correct.**
- **Your explanation may not be complete.**
- **Your explanation may not be clearly written.**

- **You did not understand what the problem asked, OR you did not know what facts were given.**
- **You did not select a good strategy or did not apply your strategy correctly.**
- **You made an arithmetic error in your work.**
- **Your explanation was not complete or you did not write an explanation.**
- **Your explanation was not clearly written.**

- **You showed no work at all,**
  OR
  the work you showed had nothing to do with the problem.

Tips

- You will never get a score of **0** if you start the problem.
- You should always write down what you were asked to find out and what facts were given. This shows that you understood some of the problem, and attempted to solve it.
- Getting used to answering a question using a rubric may seem like a lot of work, but once you start to use it, you will see it as being very helpful. You should practice using a rubric at school and at home.

2. What Is a Rubric?

Let's do an open-ended math question. See how the **5-step plan** fits the rubric to help **YOU** get a **4** on your next answer!

Remember the **5-step plan**:

1. Read and Think • **2. Select a Strategy**
3. Solve • **4. Write/Explain** • **5. Reflect**

1. Read and Think

Read the **problem** carefully.

Modeled Problem

Kelly and Sam are picking apples. Sam collected 6 bags with 4 apples in **each** bag. Kelly collected 4 bags. She had 6 apples in each bag. Who has **more** apples?

Keywords: each, more

What **question** are you asked?

- **The question tells you what it is you want to find out.**
- **The question is, "Who has more apples?"**

What are the **keywords**?

- **each** **to consider individually**
- **more** **greater in number or amount**

What **facts** are you given?

- **Every problem has facts, data, or information. Facts help you answer the question.**

In this problem, the **facts** are:

- **Kelly collected 4 bags with 6 apples in each.**
- **Sam collected 6 bags with 4 apples in each.**

2. Select a Strategy

In order to solve a problem, you need to use a **strategy**.

There are many strategies you can use. *Chapter 1, pages 8–9* shows some strategies you might use. You may also choose to use one of your own.

Let's look at how two students chose different strategies to solve this problem.

First Solution

First we'll show what Tom did. This involves using a strategy, **Writing a Number Sentence**.

3. Solve

Write a Number Sentence:

<u>Kelly's apples</u>
4 bags with 6 apples in each bag

$4 \times 6 = 24$

<u>Sam's apples</u>
6 bags with 4 apples in each bag

$6 \times 4 = 24$

Kelly and Sam have the same number of apples.

Second Solution

Second, we'll review what Carla did using a strategy called **Acting It Out**.

Act It Out:

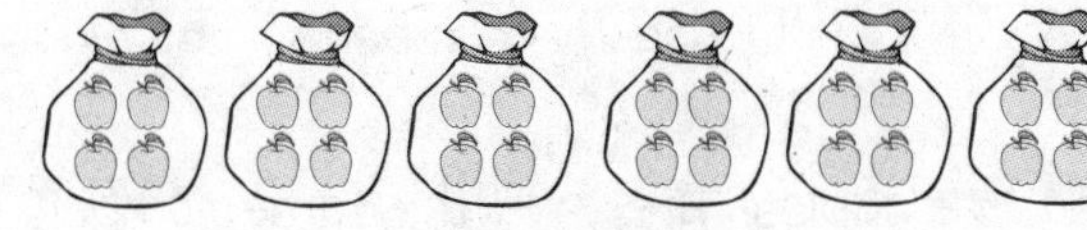

Sam's Apples

Kelly's Apples

Sam has more apples.

Tip

- Many problems can be solved using different strategies. As long as your choice leads to a correct answer and a correct explanation, it is a good choice!

First Solution

4. Write/Explain

You must give a written explanation of how you solved the problem and what you were thinking. Clearly explain what you did and why you did it. Do not leave out any steps.

I multiplied 4 x 6 = 24 to find the number of apples Kelly had.
I multiplied 6 x 4 = 24 to find the number of apples Sam had.
They each had 24 apples. They had the same number.

5. Reflect

Tom reviewed his work by checking it against the rubric. He answered the problem being asked, chose a good strategy, and used it correctly. His arithmetic was correct. He wrote a complete and clear answer to explain why he chose the strategy and how he solved the problem. Finally, he labeled his work.

Score:

Tom gets a score of **4**.

Second Solution

4. Write/Explain

Since Kelly collected 4 bags with 6 apples in each, I took 4 cups and put 6 chips in each cup. Then I added 6 + 6 + 6 + 6 and got 24.
Since Sam collected 6 bags with 4 apples in each, I took 6 cups and put 4 chips in each cup. Then I added 4 + 4 + 4 + 4 + 4 + 4 and got 26.
This shows that Sam has more apples than Kelly.

5. Reflect

Carla reviewed her work by checking it against the rubric. She answered the problem being asked. Carla chose a good strategy and used it correctly. She wrote a complete and clear answer and labeled her work. However, Carla did not catch her arithmetic error. Her addition total for Sam's apples is incorrect.

Score:

Carla gets a score of **3**.

She could improve her work to get a **4**, by adding correctly and getting 24 apples for the number Sam had.

Using the Rubric

Whenever you solve an open-ended math question, you can use the rubric on *page 13* as a guide. Use the list in the score of **4** box as a checklist. This will remind you what to include in your answer for the highest score possible.

Reviewing Your and a Partner's Work

After you finish solving the question, **self-assess**. This means that you should use the whole rubric to review your work and to score it. How well did you do? If you need to raise your score, take the time to do so.

You may also **peer-assess**. Swap your work with a partner. Use the rubric to score each other's solution. Now talk about the different answers and the scores that were given. What might seem clear to you may not be clear to your friend. Partners can help each other learn what should be improved. You can discuss different ways to solve the same problem. The more you talk about your mathematics, the more you will understand how to improve your work.

2-Step Decision-Making Process

Some students find it easier before scoring with a rubric, to first use the **2-Step Decision-Making Process** as seen in the next column. It helps decide if you or your partner's answer is a **3 or 4** or a **1 or 2**.

2-Step Decision-Making Process

Before you use a rubric, use a **2-Step Decision-Making Process**. This will give you a jump on scoring your work or your partner's work.

Decide if your work is:

- ***acceptable*** **(3 or 4)**

  or
- ***unacceptable*** **(1 or 2)**.

If your work is ***acceptable***, decide if it is:

- **full and complete (4)**

  or
- **nearly full and complete, but not perfect (3)**.

If your work is ***unacceptable***, decide if it shows:

- **limited or only some understanding (2)**

  or
- **little or no understanding (1)**.

A **(0)** is **no attempt**.

Solve questions using the rubric as a guide. You will see improvement in your answers and ability. In time you may no longer need a rubric to guide you. The information that must be included in your answer will become very familiar. Then, you will only need the rubric to score your answer.

3. How to Answer an Open-Ended Math Question

We know what an **open-ended math problem** is. We know how to solve it and how it will be scored. Now let's take a problem and solve it together. Then we will see how other students answered it.

We will use the **rubric** to see how well they did. Then we will talk about how they could **improve** their answers.

Modeled Problem

Mary's front lawn is in the **shape** of a **square**. A **side** is 20 feet long. Her father wants to put a **border** of rose bushes around the lawn. The rose bushes will be planted 4 feet **apart**, and there will be a rose bush in each **corner**. How **many** rose bushes will the family plant?

Keywords: shape, square, side, border, apart, corner, many

1. Read and Think

What **question** are we asked?

- **How many rose bushes are needed?**

What are the **keywords**?

- **shape, square, side, border, apart, corner, many**

What **facts** are we given?

- **The lawn is a square with 20-foot sides.**
- **The rose bushes will be planted 4 feet apart.**
- **There will be one rose bush in each corner.**

What is **going on**?

- **We know that the lawn is a square. This means that all four sides are 20 feet long. The family is going to plant rose bushes around the border of the lawn.**

2. Select a Strategy

Let's **Draw a Picture** to see what the lawn and its border look like. Draw a Picture is one of our strategies. The lawn is a square, 20 feet on each side.

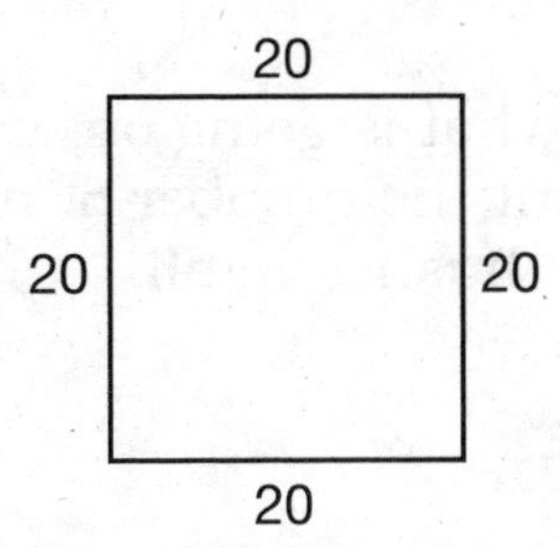

Let's put the corner rose bushes into our drawing.

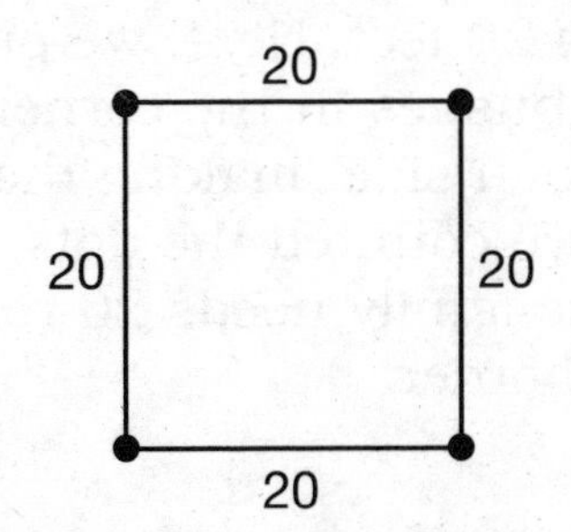

3. How to Answer...

Now let's put the rest of the rose bushes along each side of the border. They must be 4 feet apart.

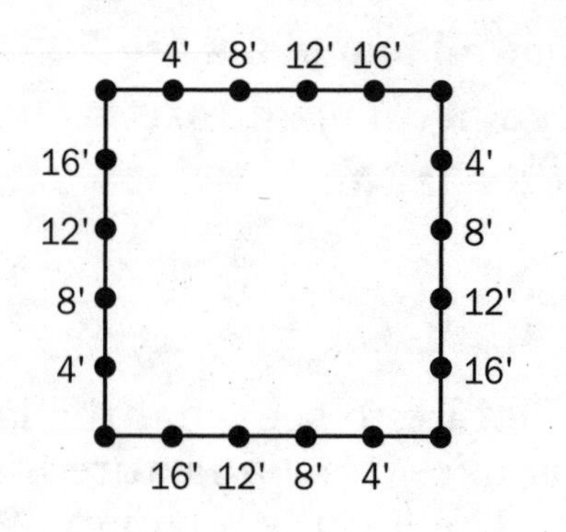

3. Solve

Now that we see what is going on, we can find the answer. Let's count the number of rose bushes. There are 20 rose bushes in all.

4. Write/Explain

We started by Drawing a Picture of the square lawn. Since all sides of a square are equal, each of the four sides is 20 feet. Next, we put dots to show the rose bushes in the corners. Then we put dots on each side, making them 4 feet apart. Next we counted the dots. There were 20 dots. The family needs 20 rose bushes to make the border.

5. Reflect

We reviewed our work. There is one rose bush in each corner. The rose bushes are all 4 feet apart. The drawing is clear and right.

Score:

This solution would earn a perfect **4** on our rubric.

- **We showed that we knew what was asked and what information was given, including facts and keywords.**
- **We chose a good strategy and applied it correctly.**
- **We did the arithmetic and got the correct answer.**
- **We labeled our work.**
- **We included all of our steps.**
- **We wrote a good explanation of how we used our strategy.**
- **We clearly explained what we did and why.**

Now let's look at some answers that were done by other students.

Michael's Paper

4' 8' 12' 16'

Each side has 6 rose bushes. There are 4 sides to a square, so there are 6 x 4 = 24 rose bushes.

Write/Explain: Here is how I found my answer. I drew one side of the square. Then I put a rose bush in each corner. Then I put the rose bushes 4 feet apart, and counted 6 rose bushes on the side. Since there are 4 sides to the square, and each side is equal, I multiplied 4 X 6 = 24 to get my answer.

Let's use our rubric to see how well Michael did.

- Did he show that he knew what the problem asked? **Yes.**
- Did he know what the keywords were? **Yes.**
- Did he show that he knew what facts were given? **Yes.**
- Did he name and use the correct strategy? **Yes. He Drew a Picture, but it was only a part of the drawing. He only showed one side of the square.**
- Was his mathematics correct? **No. Michael only drew part of the drawing so he did not see that his drawing would give him the wrong answer.**
- Did he label his work? **Yes.**
- Was his answer correct? **No. He got 24 bushes instead of 20.**
- Were all of his steps included? **No. He needed to finish his drawing for this step to be complete.**
- Did he write a good, clear explanation of his work? **Yes.**

Score:

Michael would get a **3** on our rubric.

He did not use the strategy completely. He only drew one side of the square. Michael should have drawn all four sides of the square. If he had done that, he would have counted only 20 rose bushes. He did not see that he had counted the bushes in each corner twice.

To get a 4, Michael should have gone back to check his work. He would have seen that he made a mistake in counting.

3. How to Answer...

Robin's Paper

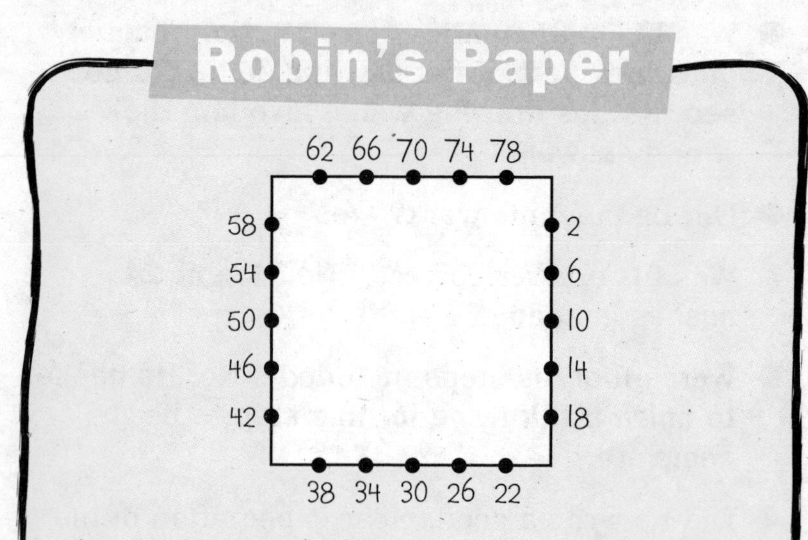

They need 20 rose bushes.

Write/Explain: I drew a picture of the square lawn and put in the rose bushes 4 feet apart. Then I counted the rose bushes. They need 20 rose bushes.

- Did she show that she knew what the problem asked? **No.**
- Did she know what the keywords were? **No. She missed 'corner.'**
- Did she show that she knew what facts were given? **She missed that there was a rosebush in each corner.**
- Did she name and use the correct strategy? **Yes.**
- Was her mathematics correct? **Yes.**
- Did she label her work? **Yes.**
- Was her answer correct? **Yes.**
- Were all of her steps included? **Yes.**
- Did she write a good, clear explanation of her work? **No.**

Score

Robin would receive a **3** on our rubric.

To get a 4, she would have to fix her drawing by drawing the rosebushes, starting in each corner.

Martha's Paper

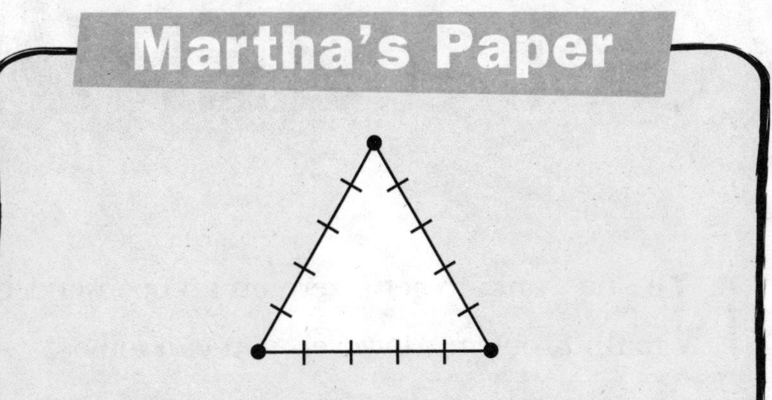

They need 15 rose bushes.

Write/Explain: I drew a picture of the lawn and put in the rose bushes 4 feet apart with one in each corner. Then I counted the rose bushes. They need 15 rose bushes.

- Did she show what the problem asked? **Yes.**
- Did she know what the keywords were? **No. She missed 'square.'**
- Did she show that she knew what facts were given? **No. She drew a triangle.**
- Did she name and use the correct strategy? **Yes.**
- Was her mathematics correct? **No. Drawing a triangle made her math wrong.**
- Did she label her work? **Yes.**
- Was her answer correct? **No.**
- Were all of her steps included? **Yes.**
- Did she write a clear explanation? **Yes.**

Score

Martha would receive a **2** on our rubric.

To get a 4, Martha needed to draw a square and recount the correct number of rose bushes.

Gary's Paper

4 x 20 = 80

Write/Explain: They need 80 rose bushes.

Score

Gary would receive a **1** on our rubric.

Gary showed a square has 4 sides. He did not write an explanation of his work, including the strategy used. The work Gary showed had nothing to do with the problem.

To get a 4, Gary needs to understand this problem and then start from the beginning.

4. How NOT to Get a ZERO!

No one wants to get a zero on an open-ended math problem. However, you can almost always get some points. The only person who gets a **0** is the person who **leaves the paper blank** or who **writes something that doesn't have anything to do with the problem**. Let's see how we can start by scoring a **1 or 2** on our work, and then bring it up to a **3 or 4**.

And remember:

1. **Read and Think**
2. **Select a Strategy**
3. **Solve**
4. **Write/Explain**
5. **Reflect**

How to Get a 1 or 2!

Here is how to get **some credit** on an open-ended math question.

1. **Read** the question. Then **reread** it.

   Ask: "What are the **keywords** to help you solve the problem?"

   Finish the sentence: **"The keyword/s are**

   ______________________________________."

You will get credit for listing the keywords.

2. **Understand** the problem. **Repeat the story** of the problem in your own words.

   Ask: "What am I **being asked** to do? What do I **need to find**?"

   Finish the sentence: **"I need to find**

   ______________________________________."

You will get credit for listing what was asked.

3. **Find the facts** in the problem.

   Ask: "What does the problem tell me? What do I **know**?"

   Finish the sentence: **"The things I know are**

   ______________________________________."

You will get credit for writing the facts.

4. Figure out what **strategy** you will use to help you solve the problem.

   Ask: "What **can help** me to find what I need to know?

   Finish the sentence: **"The strategy I will use is**

   ____

You will get credit for listing the strategy you use.

How to Get a 1 or 2!
continued on next page

- You will practice these steps as you help solve the modeled problem introduced on the following page.

Tip

- To get points right away, always begin by writing down what you are asked to find and what you are given as facts.

4. How NOT to Get a ZERO!

Let's do an open-ended math problem toegther. First, let's try to get a 1 or 2.

Modeled Problem

Mary went to lunch with her brother, Larry. She had a sandwich for $2.20 and a juice for $0.75. Larry had a pizza for $3.25 and a milk shake for $2.95. How much **more** did Larry's lunch cost than Mary's?

Keyword: more

1. Read and Think

1. Carefully **read** this question. **Reread** the question to fully understand it.
2. What are you **being asked** to find?
   - **I need to find how much more Larry's lunch cost than Mary's lunch.**

   By writing what you are being asked to find, you can get a score of a 1 or 2.

3. What are the **keywords** to help you solve the problem?
   - **more**

   By listing the keywords, you can get a score of 1 or 2.

4. What are the **facts**?
   - **Mary spent $2.20 on a sandwich.**
   - **Mary spent $0.75 on a juice.**
   - **Larry spent $3.25 on pizza.**
   - **Larry spent $2.95 on a milk shake.**

   By listing the facts, you can get a score of a 1 or 2.

2. Select a Strategy

Now you have to pick a strategy and solve the problem. There are lots of strategies from which to pick. You may choose one from *pages 8–9* or choose one of your own. Your classmate may choose a different strategy than you to solve the same problem. You may solve it differently, but you both can find the right answer. There is not only one right way.

1. What **strategy** will you use?
   - **I will use a strategy called Divide and Conquer. First I will Make a Table to**

organize all the facts in the problem. Then I will use Write Number Sentences to help me solve the problem and find the answer.

By writing what strategy you chose you can get a score of a 1 or 2.

Now, let's try to increase our score of the same problem from a 1 or 2 to a 3 or 4!.

How to Get a 3 or 4!

Remember, you can always get some credit for listing keywords and the question that is asked. You will also receive points by writing the facts that are given. Finally, credit will be given for listing the strategy you have chosen. By doing this, you will receive a score of at least a **1 or 2**. Now it is time to raise your score to a **3 or 4**. Use all this information to solve it and write how you solve it.

So here we go: Here is how to take a score of **1 or 2** and make it a **3 or 4**. We will continue using the same modeled problem.

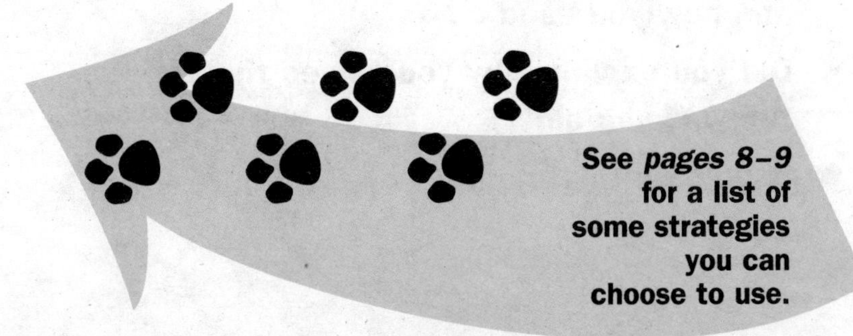

See *pages 8–9* for a list of some strategies you can choose to use.

3. Solve

Go back to read the problem. How can we find how much more Larry's lunch cost than Mary's? We use the Divide and Conquer strategy. *First*, we Make a Table to organize the data.

Name	Larry	Mary
Food	Pizza $3.25	Sandwich $2.20
Drink	Shake $2.95	Juice $0.75
TOTAL	$6.20	$2.95

Then, we Write Number Sentences to find out how much each lunch cost.

Larry's Lunch	Mary's Lunch
$3.25	$2.20
+ 2.95	+ 0.75
$6.20	$2.95

Mary's lunch cost $2.95 and Larry's lunch cost $6.20. Now we Write a Number sentence to subtract.

$6.20
– 2.95
$3.25

Larry's lunch cost $3.25 more than Mary's lunch.

Tip

- We're going to use a lot of strategies in this book.
- An open-ended problem can be solved in more than one way. And if your way works out and gives the correct answer, then you are right!

4. How NOT to Get a ZERO!

Making the Table helped us see how the facts fit together. **Writing Number Sentences** helped us solve the problem.

Always remember to **label** your work. Use **units** or a sentence that explains what you found.

4. Write/Explain

The person marking your paper does not know what you were thinking. You must explain why you chose the strategy. You should explain how you solved the problem. Your work should be labeled. Don't leave out any steps. Be sure to reread your writing, making sure your work is clear and complete.

Mary and Larry are eating lunch. We had to find how much more Larry's lunch cost than Mary's. We used the Divide and Conquer Strategy. First we Made a Table and labeled it to organize and make sense of all the facts in the problem. These facts included the prices of all the food Larry and Mary ate. Then we Wrote Number Sentences to find how much more Larry's lunch costs than Mary's. First we added the cost of each lunch separately. Larry's pizza and shake cost $6.20 in all. Mary's sandwich and juice cost $2.95 in all. Then we subtracted the cost of Mary's lunch from Larry's. We found out that Larry's lunch cost $ 3.25 more than Mary's lunch.

5. Reflect

Review Your Work and Improve It!

After you solve the problem, carefully review your work.

- **Did you write what the problem asked you to find?**
- **Did you list all keywords and facts?**
- **Did you list the strategy you chose to use?**

If you did these things, you will get a score of 1 or 2

- **Did you use the right strategy?**
- **Is your arithmetic right?**
- **Did you label your work?**
- **Did you write out all the steps to solving the problem?**
- **Did you explain why you chose the strategy and how you used it?**
- **Did you explain why you solved the problem the way you did?**
- **Is your writing clear?**

If you did these things, you will raise your score from a 1 or 2 to a 3 or 4.

If you do all the things we have suggested, you **CANNOT** get a **0**.

Remember, never leave your paper blank.

Working with Peers

You might want to exchange papers with a friend in class. See if your friend understands what you wrote. That's a good way to see how clearly you explained your work.

Here is a checklist for you to follow. It will make sure you have done your best job. Keep practicing what is on this list. You will improve at solving open-ended mathematics questions.

Read It and Think

- ☐ **Did I read the problem at least twice? Do I understand it?**
- ☐ **Did I write down the question being asked?**
- ☐ **Did I write down the keywords in the problem?**
- ☐ **Did I write down the facts that are given?**
- ☐ **Did I write down the strategy that I used?**
- ☐ **Did I solve the problem?**
- ☐ **Is my arithmetic correct?**
- ☐ **Did I explain how I solved the problem?**
- ☐ **Did I explain why I chose the strategy and how I used it?**
- ☐ **Did I include all the steps I took to solve it?**
- ☐ **Is my writing clear?**
- ☐ **Did I label my work?**
- ☐ **Does my answer make sense?**
- ☐ **Did I answer the exact question being asked?**

5. Number and Operations

Numbers are all around us, every day. They make up our **phone number**, tell us the **location** of where we live, let us know **how much** something costs, and on and on! We are going to learn more about numbers, explaining how you use them. We will understand using one number alone, how numbers can be used together, how they can be compared, and more.

Here is a problem that you might have to solve on a test. Let's solve it together to show what a model answer might look like. Then we can score it using a **rubric**.

Modeled Problem

Four brothers set up an aquarium. They spent $13 for the tank, $9 for a pump, $8 for fish, and $2 for food. They **shared** the expenses **equally**. How much did each of them spend?

Keywords: shared, equally

1. Read and Think

What **question** are we asked?

- **We are asked to find out how much money each brother spent.**

What are the **keywords**?

- **shared**
- **equally**

What **facts** were we given?

- **They spent $13 for the tank.**
- **They spent $9 for the pump.**
- **They spent $8 for the fish.**
- **They spent $2 for food.**
- **The 4 brothers shared the cost equally.**

2. Select a Strategy

There are *two parts* to this problem. The first part is *addition*. The second part is *division*. We will use the **Divide and Conquer** strategy.

3. Solve

First let's add to find out how much the brothers spent to set up the aquarium.

$$\begin{array}{r} \$13 \\ 9 \\ 8 \\ +\ 2 \\ \hline \$32 \end{array}$$

Together they spent $32.

Now, let's divide to find how much each brother spent?

$32 ÷ 4 = $8

Each brother spent $8.

4. Write/Explain

To solve this problem, we had to find out how much the brothers spent all together. So, we added to find the cost to set up the aquarium. It cost $32. Then, we divided $32 into 4 equal parts. They paid $8 each.

5. Reflect

Let's review our work and answer.

- Did we show that we knew what the problem asked for? **Yes.**
- Did we know what the keywords were? **Yes.**
- Did we show that we knew what facts were given? **Yes.**
- Did we name and use the correct strategy? **Yes.**
- Was our mathematics correct? **Yes. We checked it. It was correct.**
- Did we label our work? **Yes.**
- Was our answer correct? **Yes.**
- Were all of our steps included? **Yes.**
- Did we write a good, clear explanation of our work? **Yes.**

Score

This solution would earn a **4** on our rubric. It is perfect.

Here are some **Guided Open-Ended Math Problems**.

For each problem there are **four parts**. In the **first part**, you will solve the problem with guided help. In the **second part**, you will score and correct a solution with guided help. The **third part** shows one solution that scores a perfect **4**. This solution may or may not differ from your own way. The **fourth part** has *answers* to the **first** and **second parts** so you can check your work.

Guided Problem #1

Mrs. Applebee planted 2 rows of peppers with 9 plants in each row. She also planted 3 rows of tomatoes with 6 plants in each row. How many plants did she plant all together?

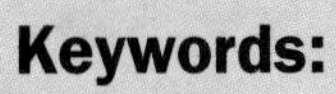

1. Try It Yourself.

Answer the questions below to get a score of **4**.

What **question** are you being asked?

____

What are the **keywords**?

____

What are the **facts** you need to solve the problem?

____

____

What **strategy** can you use to solve the problem?

____

____

Hint

Possible answers include **Divide and Conquer**, and **Draw a Picture**.

Solve the problem.

____

____

____

____

Write/Explain what you did to solve the problem.

____

____

Reflect. Review and improve your work.

____

____

2. Scott Tries It.

Scott's Paper

Keywords: all together, each

Facts: 2 rows of peppers with 9 plants and 3 rows of tomatoes with 6 plants in each row

Solve:

$3 \times 6 = 18$

$2 \times 9 = 18$

$18 = 18$

She planted the same number of pepper plants and tomato plants.

Write/Explain: I multiplied 3 rows by 6 plants in each row and got 18 tomato plants. Then I multiplied 2 rows by 9 plants in each row and got 18 pepper plants. 18 is the same as 18, so they are equal.

Score the Answer.

According to the rubric, from **1** to **3** what score would you give Scott? Explain why you gave that score.

____

____

____

____

Make it a 4! Rewrite.

____

____

____

____

____

____

3. Liu Tries It.

Remember there is often more than one way to solve a problem. Here is how Liu solved this problem.

Liu's Paper

Question: How many plants were planted?

Keywords: all together, each

Facts: 2 rows of peppers with 9 plants in each row and 3 rows of tomatoes with 6 plants in each row

Strategy: I decided to Draw a Picture.

Solve:

Peppers

= 9

= 9

9 + 9 = 18 pepper plants

Tomatoes

= 6

= 6

= 6

6 + 6 + 6 = 18 tomato plants

18 + 18 = 36

Mrs. Applebee planted 36 plants all together.

Write/Explain: I drew 2 rows of pepper plants with 9 plants in each row. There are 18 pepper plants. I drew 3 rows of tomato plants with 6 plants in each row. There are 18 tomato plants. I added the 18 pepper plants and the 18 tomato plants to get 36 plants all together.

Score: Liu's solution would earn a **4** on a test. Liu identified the question that was asked, the keywords, and the facts, labeled her work, and picked a good strategy. She explained how she used it. Then she clearly explained the steps taken to solve the problem. It is perfect!

4. Answers to Parts 1 and 2.

Guided Problem #1

Mrs. Applebee planted 2 rows of peppers with 9 plants in each row. She also planted 3 rows of tomatoes with 6 plants in each row. How many plants did she plant all together?

Keywords: each, all together

1. Try It Yourself. (pages 33–34)

Question: How many plants did Mrs. Applebee plant all together?

Facts: 2 rows of peppers with 9 plants in each row
3 rows of tomatoes with 6 plants in each row

Strategy: Divide and Conquer

Solve:
peppers: $2 \times 9 = 18$
tomatoes: $3 \times 6 = 18$
$18 + 18 = 36$

Write/Explain: I used the Divide and Conquer strategy. There are *two parts*. First, I *multiplied* to find the total number of peppers and the total number of tomatoes. Second, I *added* the products to find the total number of plants Mrs. Applebee planted. Mrs. Applebee planted 36 plants all together.

2. Scott Tries It. (page 34)

Score the Answer: I would give Scott a **3**. Scott gave the keywords and listed the facts. His math was right. Scott did not name the strategy he used. Scott did not write the question being asked in the problem. He ended up answering a different question than the one that was asked.

Make it a 4! Rewrite.

Find the total number of plants Mrs. Applebee planted:

Find the peppers: $2 \times 9 = 18$
Find the tomatoes: $3 \times 6 = 18$
Find the total number of plants: $18 + 18 = 36$

I used Number Sentences. First, I multiplied 3 rows by 6 plants in each row and got 18 tomato plants. Then I multiplied 2 rows by 9 plants in each row and got 18 pepper plants. I added $18 + 18 = 36$ to find the total number of plants Mrs. Applebee planted.

Guided Problem #2

The Browns went away to fish for 7 days. While fishing, Mrs. Brown caught 5 fish and Mr. Brown caught 3 fish. Their 3 children caught 3 fish each. Who caught more fish, the parents or the children?

Keywords: ? ?

1. Try It Yourself.

Answer the questions below to get a score of **4**.

What **question** are you being asked?

____

____

____

What are the **keywords**?

____

____

____

What are the **facts** you need to solve the problem?

____

____

____

What **strategy** can you use to solve the problem?

____

____

Hint

Possible answers include **Divide and Conquer**, **Act It Out**, and **Draw a Picture**.

Solve the problem.

Write/Explain what you did to solve the problem.

Reflect. Review and improve your work.

2. Mark Tries It.

Mark's Paper

Question: Who caught more fish: the parents or the children?

Keywords: each, more

Facts: Mrs. Brown caught 5 fish and Mr. Brown caught 3 fish. Each child caught 3 fish.

Strategy: Divide and Conquer.

Solve:

parents: $5 + 3 = 9$. The parents caught 9 fish

children: $3 \times 3 = 9$. The children caught 9 fish.

$9 = 9$

The children caught the same number of fish as their parents.

Write/Explain: I used the Divide and Conquer strategy. For the first part, I found out how many fish the parents caught all together by Writing a Number Sentence. It was 9. Then for the second part, I Wrote another Number Sentence. I found out how many fish the children caught. It was also 9. The parents and the children caught the same number of fish.

Score the Answer.

According to the rubric, from **1** to **3** what score would you give Mark? Explain why you gave that score.

Make it a 4! Rewrite.

3. Helen Tries It.

Remember, there is often more than one way to solve a problem. Here is how Helen solved this problem.

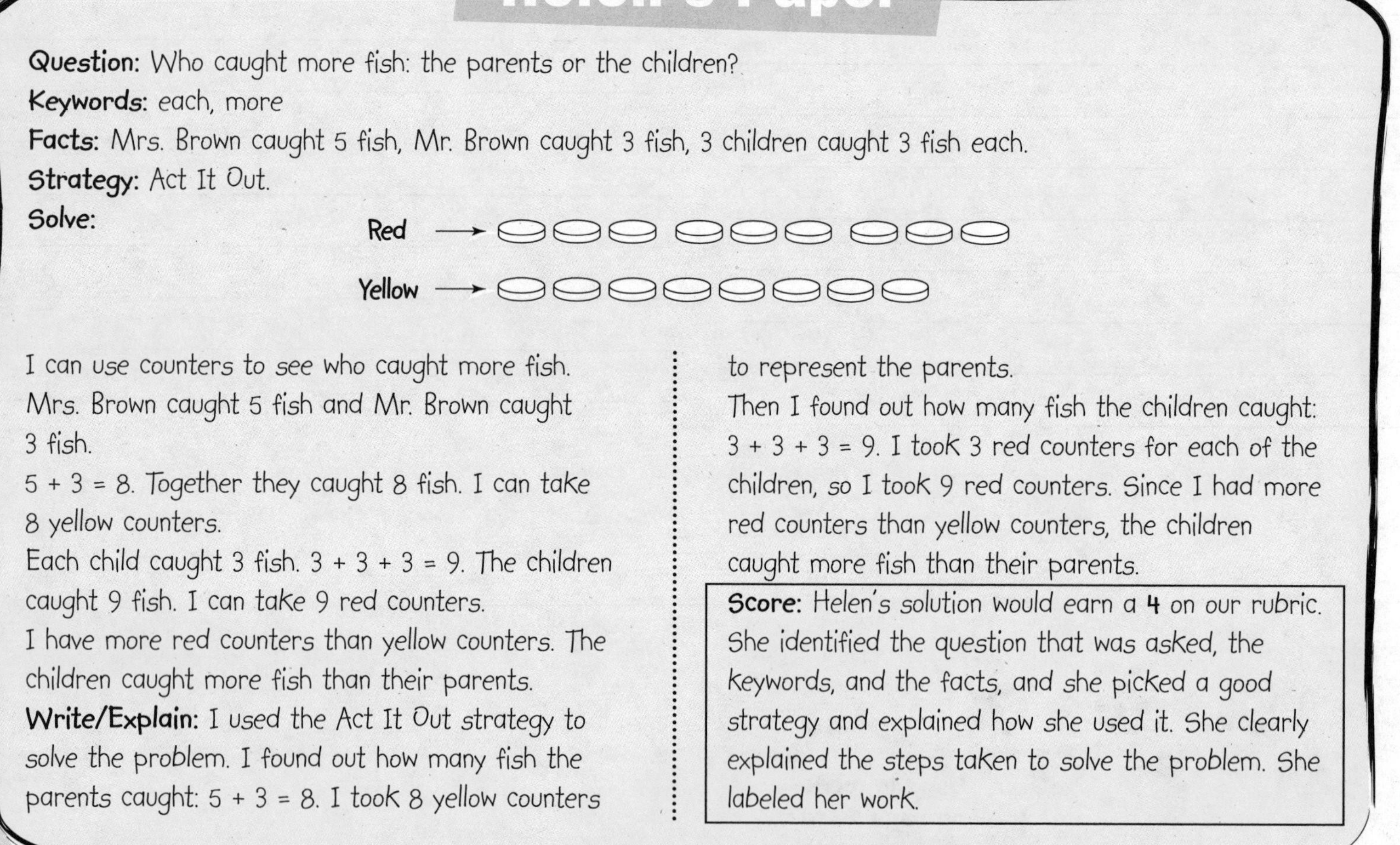

Helen's Paper

Question: Who caught more fish: the parents or the children?

Keywords: each, more

Facts: Mrs. Brown caught 5 fish, Mr. Brown caught 3 fish, 3 children caught 3 fish each.

Strategy: Act It Out.

Solve:

I can use counters to see who caught more fish.
Mrs. Brown caught 5 fish and Mr. Brown caught 3 fish.
5 + 3 = 8. Together they caught 8 fish. I can take 8 yellow counters.
Each child caught 3 fish. 3 + 3 + 3 = 9. The children caught 9 fish. I can take 9 red counters.
I have more red counters than yellow counters. The children caught more fish than their parents.

Write/Explain: I used the Act It Out strategy to solve the problem. I found out how many fish the parents caught: 5 + 3 = 8. I took 8 yellow counters to represent the parents.
Then I found out how many fish the children caught: 3 + 3 + 3 = 9. I took 3 red counters for each of the children, so I took 9 red counters. Since I had more red counters than yellow counters, the children caught more fish than their parents.

Score: Helen's solution would earn a **4** on our rubric. She identified the question that was asked, the keywords, and the facts, and she picked a good strategy and explained how she used it. She clearly explained the steps taken to solve the problem. She labeled her work.

4. Answers to Parts 1 and 2.

Guided Problem #2

The Browns went away to fish for 7 days. While fishing, Mrs. Brown caught 5 fish and Mr. Brown caught 3 fish. Their 3 children caught 3 fish each. Who caught more fish, the parents or the children?

Keywords: each, more

1. Try It Yourself. (pages 37–38)

Question: Who caught more fish: the parents or the children?

Facts: Mrs. Brown caught 5 fish.
Mr. Brown caught 3 fish.
3 children caught 3 fish each.

Strategy: Divide and Conquer

Solve: parents: $5 + 3 = 8$
children: $3 \times 3 = 9$
$8 < 9$

The children caught more fish.

Write/Explain: I used the Divide and Conquer strategy. There are *two parts*. For the first part, I *added* to find the number of fish Mr. and Mrs. Brown caught. For the second part, I *multiplied* to find the number of fish the children caught. I then compared the numbers.

2. Mark Tries It. (pages 38–39)

Score the Answer: Mark would get a **3** because his addition is wrong. He said $5 + 3 = 9$. He gave the keywords and the facts. He knew what the question asked and he picked a good strategy. He labeled his work.

Make it a 4! Rewrite.

Find the number of fish the parents caught:

$5 + 3 = 8$

Find the number of fish the children caught:

$3 \times 3 = 9$

Compare 8 and 9.

$8 < 9$, so the children caught more fish than the parents.

I added $5 + 3 = 8$ to find the number of fish the parents caught. I multiplied $3 \times 3 = 9$ to find the number of fish the children caught. I compared 8 and 9. Since 8 is less than 9, the children caught more fish than the parents.

Guided Problem #3

There are 24 balls in the gym storage room.

There are 12 footballs and 4 volleyballs.

The rest of the balls are basketballs.

How many basketballs are in the storage room?

Keywords: ? ?

1. Try It Yourself.

Answer the questions below to get a score of **4**.

What **question** are you being asked?

What are the **keywords**?

What are the **facts** you need to solve the problem?

What **strategy** can you use to solve the problem?

Hint

Possible answers include **Divide and Conquer**, **Draw a Picture**.

Solve the problem.

Write/Explain what you did to solve the problem.

Reflect. Review and improve your work.

2. Michaela Tries It.

Michaela's Paper

Question: How many balls in all?

Keywords: rest, many

Facts: 24 balls in the gym. 12 are footballs. 4 are volleyballs. The rest are basketballs.

Solve:

$$\begin{array}{r} 24 \\ 12 \\ +\ 4 \\ \hline 40 \end{array}$$

The total number of balls is 40.

Write/Explain: I added the balls, the footballs, and the volleyballs to find the total number of balls.

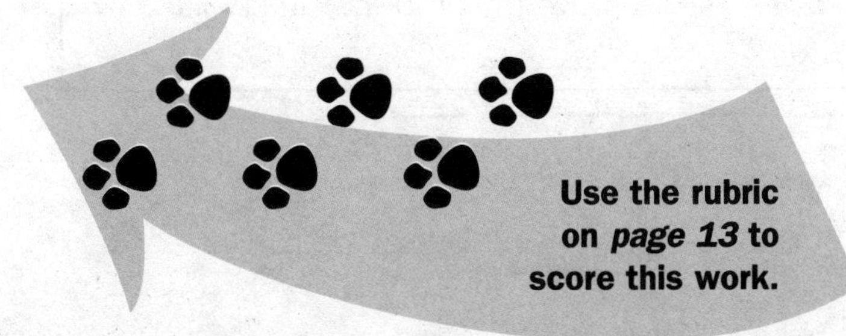

Score the Answer.

According to the rubric, from **1** to **3** what score would you give Michaela? Explain why you gave that score.

____

____

____

____

____

____

____

Make it a 4! Rewrite.

____

____

____

____

____

____

____

____

3. Ian Tries It.

Remember, there is often more than one way to solve a problem. Here is how Ian solved this problem.

Ian's Paper

Question: How many balls in all?

Keywords: rest, many

Facts: 24 balls in the gym. 12 are footballs. 4 are volleyballs. The rest are basketballs.

Strategy: I drew a picture.

Solve:

F	F	F	F	F	F
F	F	F	F	F	F
V	V	V	V	B	B
B	B	B	B	B	B

There are 8 basketballs.

Write/Explain: I drew a picture. There are 24 balls all together, so I drew 24 boxes. I put 12 footballs and 4 volleyballs in the boxes. In the remaining boxes, I put basketballs. I counted 8 basketballs.

Score: Ian earned a perfect **4** on this problem. He identified the question that was asked, the keywords, and the facts. He picked a good strategy and explained how he used it. He clearly explained the steps he used to solve the problem. He labeled his work.

4. Answers to Parts 1 and 2.

Guided Problem #3

There are 24 balls in the gym storage room.

There are 12 footballs and 4 volleyballs.

The rest of the balls are basketballs.

How many basketballs are in the storage room?

Keywords: rest, many

1. Try It Yourself. (page 42)

Question: How many basketballs are there?

Facts: There are 24 balls in all. There are 12 footballs.

There are 4 volleyballs. The rest of the balls are basketballs.

Strategy: Divide and Conquer

Solve: Find the number of footballs and volleyballs:

$12 + 4 = 16$

Find the number of basketballs: $24 - 16 = 8$

Write/Explain: I used the Divide and Conquer strategy. There are *two parts.* For the first part, I Wrote a Number Sentence. I *added* to find the number of footballs and volleyballs. For the second part, I Wrote a Number Sentence. I *subtracted* that sum from the total number of balls.

2. Michaela Tries It. (page 42)

Score the Answer: Michaela would get a **2**. She used the keywords. She labeled her answer. Her addition was correct, but she did not answer the question. She added the number of footballs and volleyballs to the total number of balls. She did not understand the question.

Make it a 4! Rewrite.

Find the number of footballs and volleyballs:

$12 + 4 = 16$

Find the number of basketballs: $24 - 16 = 8$

I added to find the number of footballs and volleyballs. I subtracted that sum from the total number of balls.

Guided Problem #4

Mildred started her homework at 5:30 p.m. She spent 45 minutes on math. Then she spent 30 minutes on social studies and 30 minutes on reading. At what time did she finish her homework?

Keywords: ? ?

1. Try It Yourself.

Answer the questions below to get a score of **4**.

What **question** are you being asked?

____

____

What are the **keywords**?

____

____

What are the **facts** you need to solve the problem?

____

____

What **strategy** can you use to solve the problem?

____

____

Hint

Possible answers include **Make a Table** and **Divide and Conquer**.

Solve the problem.

____

____

Write/Explain what you did to solve the problem.

____

____

____

____

Reflect. Review and improve your work.

____

____

____

____

2. Cindy Tries It.

Cindy's Paper

Question: When did she finish her homework?
Keywords: started, finish
Facts: She started at 5:30, spent 45 min. on math, 30 min. on social studies, 30 min. on reading.
Solve: 45 + 30 + 30 = 105 minutes on her homework
105 minutes = 1 hour and 5 minutes
5:30 p.m. + 1 hour and 5 minutes = 6:35 p.m.
Mildred finished her homework at 6:35 p.m.
Write/Explain: I added the number of minutes that Mildred spent on her homework. She spent 105 minutes. I changed 105 minutes to 1 hour 5 minutes. Then I added it to 5:30 p.m. She finished her homework at 6:35 p.m.

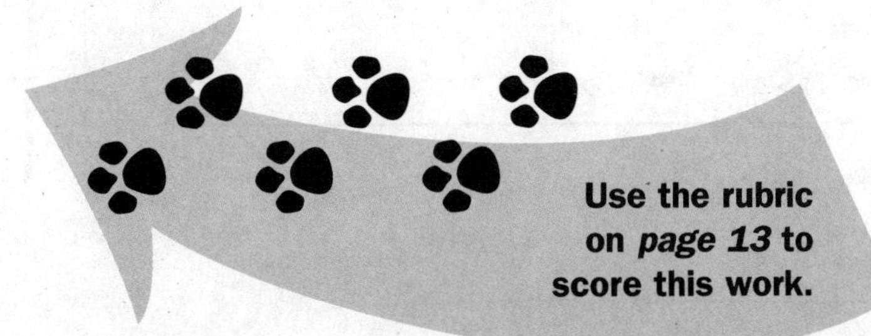
Use the rubric on *page 13* to score this work.

Score the Answer.

According to the rubric, from **1** to **3** what score would you give Cindy? Explain why you gave that score.

____

____

____

____

____

____

____

Make it a 4! Rewrite.

____

____

____

____

____

____

____

____

3. Kenny Tries It.

Remember there is often more than one way to solve a problem. Here is how Kenny solved this problem.

Kenny's Paper

Question: When did she finish her homework?

Keywords: started, finish

Facts: She started at 5:30 p.m., spent 45 min. on math, 30 min. on social studies, 30 min. on reading.

Strategy: Make a Table

Solve:

Subject	Time Began	Time Spent	Time Finished
Math	5:30 p.m.	45 min.	6:15 p.m.
Social Studies	6:15 p.m.	30 min.	6:45 p.m.
Reading	6:45 p.m.	30 min.	7:15 p.m.

Mildred finished her homework at 7:15 p.m.

Write/Explain: I made a table to organize the information. I put the time Mildred started and finished the homework for each subject. I added the minutes spent to the time she began each subject. Math began at 5:30 p.m. I figured out Social Studies began at 6:15 p.m. Then Reading began at 6:45 p.m. Reading was the last subject. So 30 minutes added to 6:45 p.m. is 7:15 p.m. The table showed that she finished at 7:15 p.m.

Score: Kenny earned a perfect **4** on our rubric. He identified the question that was asked, the keywords, and the facts. He picked a good strategy of making a table and explained how he used it. He labeled his work. He clearly explained the steps he used to solve the problem.

4. Answers to Parts 1 and 2.

Guided Problem #4

Mildred started her homework at 5:30 p.m. She spent 45 minutes on math. Then she spent 30 minutes on social studies and 30 minutes on reading. At what time did she finish her homework?

Keywords: started, finish

1. Try It Yourself. (page 46)

Question: When did Mildred finish her homework?

Facts: Mildred started at 5:30 p.m. She spent 45 minutes on math.

She spent 30 minutes on social studies. She spent 30 minutes on reading.

Strategy: Divide and Conquer

Solve: Add to find how long she spent doing her homework: 45 + 30 + 30 = 105 minutes.

Convert 105 minutes to 1 hour 45 minutes.

Count on from 5:30: 1 hour is 6:30, and 45 minutes is 7:15.

Write/Explain: I used the Divide and Conquer strategy. There are *two parts*. For the *first* part, I used a Number Sentence. I *added* the minutes Mildred spent doing homework. For the *second* part, I used a Number Sentence. I then counted on from 5:30 p.m. to find what time she finished.

2. Cindy Tries It. (page 47)

Score the Answer: Cindy would get a **3**. She found the keywords, and the facts, answered the question that was asked, and labeled her work. She should have named the strategy she used. Her error was in changing 105 min. to 1 h. 5 min.

Make it a 4! Rewrite.

45 + 30 + 30 = 105 minutes on her homework

105 minutes = 1 hour and 45 minutes

5:30 p.m. + 1 hour and 45 minutes = 7:15 p.m.

Mildred finished her homework at 7:15 p.m.

Guided Problem #5

Maria bought 3 packages of crackers that cost $0.75 each. She received two $1 bills and three quarters in change from the grocer. How much money did she give the grocer?

Keywords: 

1. Try It Yourself.

Answer the questions below to get a score of **4**.

What **question** are you being asked?

____

____

What are the **keywords**?

____

____

What are the **facts** you need to solve the problem?

____

____

What **strategy** can you use to solve the problem?

____

____

Hint

Possible answers include **Divide and Conquer**, **Draw a Picture**.

Solve the problem.

____

____

Write/Explain what you did to solve the problem.

____

____

____

Reflect. Review and improve your work.

____

____

____

____

2. Danielle Tries It.

Danielle's Paper

3 x \$0.75 = \$2.15

\$2.00 + 3 x \$0.25 = \$2.00 + \$0.65 = \$2.65

\$2.15 + \$2.65 = \$4.80

Maria gave the grocer \$4.80.

Write/Explain: I found the cost of 3 packages of crackers. They cost \$2.15. Next, I found the amount of change she received. She received \$2.65. Then I added the cost and the change, and found that Maria gave the grocer \$4.80.

Use the rubric on *page 13* to score this work.

Score the Answer.

According to the rubric, from **1** to **3** what score would you give Cindy? Explain why you gave that score.

____

____

____

____

____

____

____

Make it a 4! Rewrite.

____

____

____

____

____

____

____

____

3. Ben Tries It.

Remember, there is often more than one way to solve a problem. Here is how Ben solved this problem.

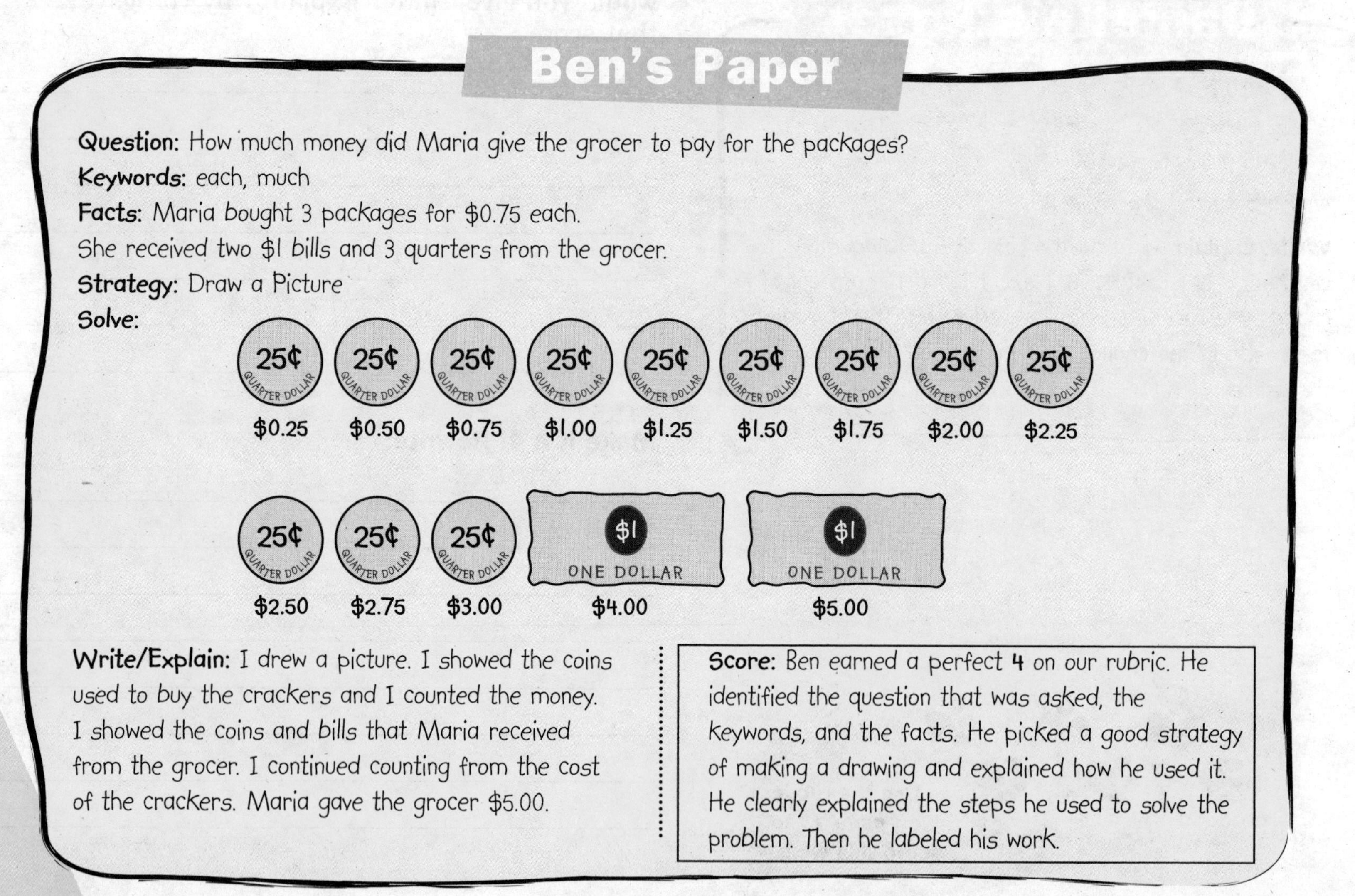

Ben's Paper

Question: How much money did Maria give the grocer to pay for the packages?

Keywords: each, much

Facts: Maria bought 3 packages for $0.75 each.

She received two $1 bills and 3 quarters from the grocer.

Strategy: Draw a Picture

Solve:

Write/Explain: I drew a picture. I showed the coins used to buy the crackers and I counted the money. I showed the coins and bills that Maria received from the grocer. I continued counting from the cost of the crackers. Maria gave the grocer $5.00.

Score: Ben earned a perfect 4 on our rubric. He identified the question that was asked, the keywords, and the facts. He picked a good strategy of making a drawing and explained how he used it. He clearly explained the steps he used to solve the problem. Then he labeled his work.

4. Answers to Parts 1 and 2.

Guided Problem #5

Maria bought 3 packages of crackers that cost $0.75 each. She received two $1 bills and three quarters in change from the grocer. How much money did she give the grocer?

Keywords: each, much

1. Try It Yourself. (page 50)

Question: How much money did Maria give the grocer?

Facts: Maria bought 3 packages for $0.75 each.

She received two $1 bills and 3 quarters from the grocer.

Strategy: Number Sentences

Solve: 3 × $0.75 = $2.25. This is the cost of the three packages of crackers.

3 × $0.25 = $0.75. This is the amount of change she received in quarters.

2 × $1.00 = $2.00. This is the amount of change she received in dollar bills.

$2.25 + $0.75 + $2.00 = $5.00

Maria gave the grocer $5.00.

Write/Explain: I used Number Sentences to help me solve this problem. I multiplied to find out how much Maria spent for the crackers. I then found the value of 3 quarters. I added $0.75 to $2.00 = $2.75, to find out how much change Maria received from the grocer. I added the change to the cost of the crackers: $2.25 + $2.75 = $5.00. Maria gave the grocer $5.

2. Danielle Tries It. (page 51)

Score the Answer: Danielle would get a **1**. She did not give the keywords, facts, strategy used, or question asked. Her answer is also incorrect.

Make it a 4! Rewrite.

Keywords: each, much

Facts: Maria bought 3 packages for $0.75 each. She received two $1 bills and 3 quarters from the grocer.

Question: How much money did she give the grocer?

Solve: I would Write a Number Sentence.

The cost of the crackers: 3 × $0.75 = $2.25

The change she received: 2 × $1.00 = $2.00 and 3 × $0.25 = $0.75; $0.75 + $2.00 = $2.75

$2.25 + $2.75 = $5.00

Maria gave the grocer $5.00.

Quiz Problems

Here are some problems for you to try. Keep your **rubric** handy while you solve the problem. Let's see if you can score a **4**.

1. Four classes are going to the science museum. The third-grade bus had 24 students from Ms. Wiley's class and 19 students from Ms. Alan's class. The fourth-grade bus had 27 students from Mr. Field's class and 21 students from Ms. Cooper's class. Which grade had more students at the museum? How many more?

2. Mrs. Roberts baked 24 cookies for her 2 children to share with their friends. Joy received 8 cookies and Lynn received the rest of the cookies. Which girl was given more cookies? How many more?

3. While at the shore, Helen collected 12 shells and Alex collected 20 shells. They decided to share the shells equally. How can they make this happen?

4. Four friends went apple picking. Amy picked 16 apples, Nancy picked 17 apples, Danny picked 20 apples, and Jeff picked 27 apples. They decided to share the apples equally. How many apples did each child get?

5. Jackie has 5 one-dollar bills, 3 quarters, and 4 dimes. She bought a piece of pie for $2.75 and a glass of milk for $1.25. How much money will she have left after paying for her food?

6. There are 34 students in the chorus. They are going to perform at a senior citizens' home. The students will be driven there by some of the parents. If each parent's car can hold 4 students, how many cars are needed?

7. Frank is 3 years younger than Ralph. Ralph is 4 years older than Mischa. Mischa is 6 years old. How old is Frank?

6. Algebra

In this chapter, we are going to look at some basic ideas of algebra. Algebra is about discovering **patterns** and finding **unknowns**. It makes you think about **number relationships** and **equations**, such as what is equal to what. You will also focus on your thoughts that lead to your best guess when solving problems.

In algebra, we use **symbols** such as a box or a letter to take the place of a number. The modeled problem here will show how this is done. Let's solve it together to show what a model answer might look like. Then we can score it using a rubric.

Modeled Problem

In 3 **years**, Sally will be 12 years old. How old is Sally **now**?

Keywords: now, years

1. Read and Think

What is the **question** we are asked?

- **We are asked to find out how old Sally is now.**

What are the **keywords**?

- **now**
- **years**

What **facts** were we given?

- **In 3 years Sally will be 12 years old.**

2. Select a Strategy

To solve this problem, we are going to **Write a Number Sentence**.

3. Solve

We can write

$3 + s = 12$.

The variable s represents how old Sally is now.

Let's write a related number sentence:

$12 - 3 = s$.

Since $12 - 3 = 9$, Sally is now 9 years old.

We labeled our work.

4. Write/Explain

We Wrote a Number Sentence to show Sally's age in 3 years.

Then we used a related number sentence to find how old Sally is now.

5. Reflect

Let's review our work and answer.

- Did we show that we knew what the problem asked? **Yes.**
- Did we know what the keywords were? **Yes.**
- Did we show that we knew what facts were given? **Yes.**
- Did we name and use the correct strategy? **Yes.**
- Was our mathematics correct? **Yes. We checked it. It was correct.**
- Did we label our work? **Yes.**
- Was our answer correct? **Yes.**
- Were all of our steps included? **Yes.**
- Did we write a good, clear explanation of our work? **Yes.**

Score

This solution would receive a **4** on our rubric.

Here are some **Guided Open-Ended Math Problems**.

For each problem there are **four parts**. In the **first part**, you will solve the problem with guided help. In the **second part**, you will score and correct a solution with guided help. The **third part** shows one solution that scores a perfect **4**. This solution may or may not differ from your way. The **fourth part** has *answers* to the **first** and **second parts** so you can check your work.

Guided Problem #1

Marcy collects dolls. Her father gave her 3 dolls. She now has 28 dolls. How many dolls did she have before her father gave her the new ones?

Keywords: ? ?

1. Try It Yourself.

Answer the questions below to get a score of **4**.

What **question** are you being asked?

____

What are the **keywords**?

____

6. Algebra

What are the **facts** you need to solve the problem?

What **strategy** can you use to solve the problem?

Hint

Possible answers include **Divide and Conquer**, **Draw a Picture**.

Solve the problem.

Write/Explain what you did to solve the problem.

Reflect. Review and improve your work.

2. Lonnie Tries It.

Lonnie's Paper

Question: How many dolls did Marcy have?

28 + 3 = 31

Keywords: before, many

Facts: Marcy got 3 dolls from her father. She now has 28 dolls.

Write/Explain: I Wrote a Number Sentence. I added the number of dolls that Marcy just got to the number she has now.

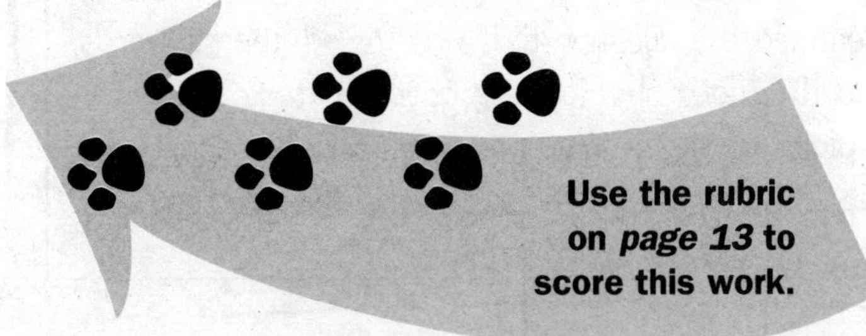

Use the rubric on *page 13* to score this work.

Score the Answer.

According to the rubric, from **1** to **3**, what score would you give Lonnie? Explain why you gave that score.

____

____

____

____

____

____

Make it a 4! Rewrite.

____

____

____

____

____

____

____

____

____

3. Jessie Tries It.

Remember, there is often more than one way to solve a problem. Here is how Jesse solved this problem.

Jesse's Paper

Question: How many dolls did Marcy have before getting the three new dolls?

Keywords: before, many

Facts: Marcy got 3 dolls from her father. She now has 28 dolls.

Strategy: Draw a Picture

Solve: I can find how many dolls Marcy had by drawing 28 boxes. I can place an X in 3 boxes for the dolls that she just received.

$$28 - 3 = 25$$

Marcy has 25 dolls.

X	X	X				

Write/Explain: I used the Draw a Picture strategy. I drew 28 boxes to represent the dolls. Then I made an X in 3 of the boxes to represent the dolls that Marcy just received. The remaining 25 boxes represent the number of dolls that Marcy already had.

Score: Jesse's solution would earn a **4** on a test. He identified the question that was asked, the keywords, and the facts. He picked a good strategy. He explained how the strategy was used. He clearly explained the steps taken to solve the problem. Jess labeled his work. It is perfect!

4. Answers to Parts 1 and 2.

Guided Problem #1

Marcy collects dolls. Her father gave her 3 dolls. She now has 28 dolls. How many dolls did she have before her father gave her the new ones?

Keywords: before, many

1. Try It Yourself. (page 60)

Question: How many dolls did Marcy have before she got the new ones?

Facts: Her father gave her 3 dolls. She now has 28 dolls.

Strategy: Write a Number Sentence.

Solve: I can Write a Number Sentence.

28 – 3 = 25

Write/Explain: I Wrote a Number Sentence. I subtracted the number of dolls that Marcy received from the number of dolls that she has now. Marcy had 25 dolls.

2. Lonnie Tries It. (page 61)

Score the Answer: I would give Lonnie a **2**. He gave the keywords, listed the strategy, and listed the facts. His math was right. But Lonnie did not write the question being asked in the problem. He ended up answering a different question than the one that was asked.

Make it a 4! Rewrite.

28 – 3 = 25

I subtracted the 3 dolls that Marcy just got from her total of 28. Since 28 – 3 = 25, she had 25 dolls before getting the new ones.

Guided Problem #2

Ms. Clarke gave her class this number puzzle.

I am thinking of 2 numbers. Their sum is 12 and their difference is 6. What are the 2 numbers? Solve Ms. Clarke's puzzle. Show your work. Explain what you did.

Keywords: 

1. Try It Yourself.

Answer the questions below to get a score of **4**.

What **question** are you being asked?

____

What are the **keywords**?

____

____

What are the **facts** you need to solve the problem?

____

____

What **strategy** can you use to solve the problem?

____

____

Hint

Possible answers include **Guess and Check** and **Use Logical Thinking**.

Solve the problem.

____

____

Write/Explain what you did to solve the problem.

____

____

____

____

Reflect. Review and improve your work.

____

____

____

2. Janet Tries It.

Janet's Paper

Question: What are the 2 numbers?
Keywords: sum, difference
Facts: Two numbers have a sum of 12 and a difference of 6.
Strategy: Write Number Sentences
Solve: Try 1 and 12

12 x 1 = 12
12 − 1 = 11

The difference is too high.
Try 2 and 6

6 x 2 = 12
10 − 2 = 8

The difference is too high.
Try 3 and 4

4 x 3 = 12
4 − 3 = 1

The difference is too low.
There is no answer to Ms. Clarke's puzzle.
Write/Explain: I wrote a pair of number sentences. I have tried all the possible numbers. No other pair of factors multiply to 12. There were only 3 pairs of numbers. Ms. Clarke was fooling us. There are no pairs of numbers that work.

Score the Answer.

According to the rubric, from **1** to **3**, what score would you give Janet? Explain why you gave that score.

____
____
____
____
____

Make it a 4! Rewrite.

____
____
____
____
____
____

3. Dave Tries It.

Remember, there is often more than one way to solve a problem. Here is how Dave solved this problem.

Dave's Paper

Question: What are the 2 numbers?

Keywords: sum, difference

Facts: Two numbers have a sum of 12 and a difference of 6.

Strategies: Logical Thinking, Make a List

Solve: Two numbers that have a sum of 12 and a difference of 6 must both be less than 12 since 12 – 12 = 0.
I can make a list and work down from 12 – 12 = 0

11 – 1 = 10

10 – 2 = 8

9 – 3 = 6

The numbers are 3 and 9.

Write/Explain: I used Logical Thinking and Made a List to find the 2 numbers. I used the greatest and least numbers that have a sum of 12 and worked down. I found that 9 – 3 = 6 and 9 + 3 = 12, so the numbers are 3 and 9.

Score: Dave's solution would earn a **4** on our rubric. He identified the question that was asked, the keywords, and the facts. He picked good strategies. He explained how he used them. Then he clearly explained the steps he took to solve the problem. And finally, he labeled his work.

4. Answers to Parts 1 and 2.

Guided Problem #2

Ms. Clarke gave her class this number puzzle.

I am thinking of 2 numbers. Their sum is 12 and their difference is 6. What are the 2 numbers?

Solve Ms. Clarke's puzzle. Show your work.

Explain what you did.

Keywords: sum, difference

1. Try It Yourself. **(page 64)**

Question: What are the 2 numbers that have a sum of 12 and a difference of 6?

Facts: Two numbers have a sum of 12. The same 2 numbers have a difference of 6.

Strategies: Write Number Sentences, Guess and Test

Solve:

Try 1 and 11	11 + 1 = 12	11 − 1 = 10	The difference is too high.
Try 2 and 10	10 + 2 = 12	10 − 2 = 8	The difference is too high.
Try 3 and 9	9 + 3 = 12	9 − 3 = 6	Yes.

Try 1 and 11.

$11 + 1 = 12$
$11 - 1 = 10$

The difference is too high.

Try 2 and 10.

$10 + 2 = 12$
$10 - 2 = 8$

The difference is too high.

Try 3 and 9.

$9 + 3 = 12$
$9 - 3 = 6$

Yes.

The numbers are 3 and 9.

Write/Explain: I Wrote 2 Number Sentences that were in the puzzle. I used the Guess and Test strategy. It took me 3 guesses to find the right pair of numbers. I checked my arithmetic and it was right. I answered the question.

2. Janet Tries It. (page 65)

Score the Answer: Janet would get a **2**. She gave the keywords and the facts. She picked a good strategy. But she was unable to find the answer. She was unable because she confused product with sum.

Make it a 4! Rewrite.

$9 + 3 = 12$ and $9 - 3 = 6$

The sum is equal to 12 and the difference is equal to 6. The 2 numbers are 3 and 9.

Guided Problem #3

Mrs. Bradley brought 24 cupcakes for her students to eat. There were twice as many chocolate cupcakes as vanilla ones. How many of each cupcake did Mrs. Bradley bring? Show how you found your answer.

Keywords: ? ?

1. Try It Yourself.

Answer the questions below to get a score of **4**.

What **question** are you being asked?

___

What are the **keywords**?

___

What are the **facts** you need to solve the problem?

___

___

What **strategy** can you use to solve the problem?

___

___

Hint

Possible answers include **Guess and Check** and **Draw a Picture**.

Solve the problem.

___

___

___

___

Write/Explain what you did to solve the problem.

___

___

Reflect. Review and improve your own.

___

___

2. Terrance Tries It.

Terrance's Paper

Question: How many of each cupcake are there?

Facts: There are 24 cupcakes in all. There are twice as many chocolate cupcakes as vanilla cupcakes.

Solve: Let V = the number of vanilla cupcakes

Let C = the number of chocolate cupcakes

Then V + C = 24

This tells me to find two numbers that add up to 24. There are lots of them, such as:

1 and 23, 2 and 22, 3 and 21, 4 and 20, and so on.

Write/Explain: Mrs. Bradley could have brought a lot of different combinations. I can't tell which way she did it.

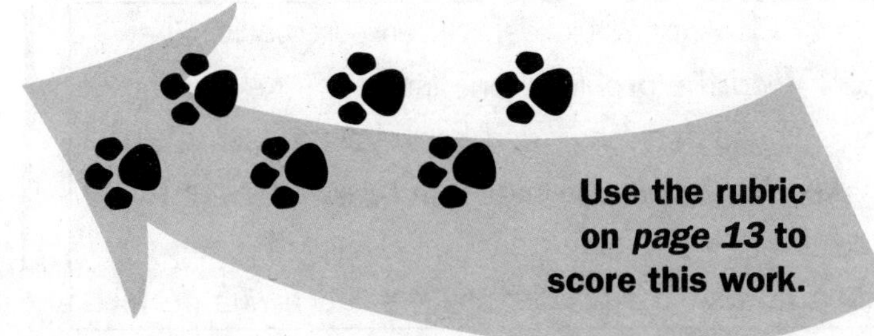

Use the rubric on *page 13* to score this work.

Score the Answer.

According to the rubric, from **1** to **3**, what score would you give Terrance? Explain why you gave that score.

Make it a 4! Rewrite.

3. Christine Tries It.

Remember, there is often more than one way to solve a problem. Here is how Christine solved this problem.

Christine's Paper

Question: How many of each cupcake are there?

Keywords: twice, many

Facts: There are 24 cupcakes in all. There are twice as many chocolate cupcakes as vanilla cupcakes.

Strategies: Guess and Test, Make a Table

Solve:

Try 12 and 6	12 ÷ 6 = 2	12 + 6 = 18	The sum is too low.
Try 14 and 7	14 ÷ 7 = 2	14 + 7 = 21	The sum is too low.
Try 16 and 8	16 ÷ 8 = 2	16 + 8 = 24	Yes.

Try 12 and 6. 12 ÷ 6 = 2
12 + 6 = 18 The sum is too low.
Try 14 and 7. 14 ÷ 7 = 2
14 + 7 = 21 The sum is too low.
Try 16 and 8. 16 ÷ 8 = 2
16 + 8 = 24 Yes.

The numbers are 16 chocolate and 8 vanilla.

Write/Explain: I used the Guess and Test strategy to find the answer. I Made a Table to keep my information organized. I chose 2 numbers that would give a quotient of 2 when I divided the greater number by the lesser one. Then I added the addends until I found a pair that had a sum of 24.

Score: Christine would get a **4** on our rubric. She understood the problem. She listed the key words, chose a good strategy, and knew how to use it. She answered the question that was being asked and labeled her work. She clearly explained the steps taken to solve the problem. Her solution was perfect.

4. Answers to Parts 1 and 2.

Guided Problem #3

Mrs. Bradley brought 24 cupcakes for her students to eat. There were twice as many chocolate cupcakes as vanilla ones. How many of each cupcake did Mrs. Bradley bring? Show how you found your answer.

Keywords: twice, many

1. Try It Yourself. (page 68)

Question: How many of each type of cupcake did she bring?

Facts: There are 24 cupcakes. There are twice as many chocolate as vanilla cupcakes.

Strategy: Draw a Picture.

Solve: There are 24 cupcakes. I Drew a Picture. For every 2 chocolate there is 1 vanilla.

C C V C C V
C C V C C V
C C V C C V
C C V C C V

Write/Explain: There are twice as many chocolate cupcakes as vanilla ones, so I separated them into groups of three and made 2 chocolate and 1 vanilla. I counted 16 chocolate ones and 8 vanilla ones.

2. Terrance Tries It. (page 69)

Score the Answer: Terrance would get a **1** because he did not attempt to answer the question. He knew the facts and what was asked, but did not know how to solve the problem. He also forgot to give the keywords.

Make it a 4! Rewrite.

The keywords are *twice* and *many*. There are 24 cupcakes. So use V for vanilla and C for chocolate. Since there are 2 chocolate for each vanilla, write 2 C for every V.

C C V C C V C C V C C V
C C V C C V C C V C C V

There are 16 chocolate and 8 vanilla cupcakes.

Guided Problem #4

Alex and Belinda started the race at 1:00 p.m. It took Alex 10 seconds longer than Belinda to finish a race. Belinda took 5 seconds less than Carol to finish. Carol finished the race in 48 seconds. How long did it take Alex to finish the race?

Keywords: ? ?

1. Try It Yourself.

Answer the questions below to get a score of **4**.

What **question** are you being asked to find?

____

What are the **keywords?**

____

____

What are the **facts** you need to solve the problem?

____

____

____

What **strategy** can you use to solve the problem?

____

____

Possible answers include **Work Backward** and **Write a Number Sentence**.

Solve the problem.

Write/Explain what you did to solve the problem.

Reflect. Review and improve your work.

2. Miguel Tries It.

Miguel's Paper

Question: How long did it take Alex?
Keywords: longer, less
Facts: Alex took 10 s. longer than Belinda. Belinda took 5 s. less than Carol. It took Carol 48 s.
Strategy: Work Backward
Solve: Let A = the time it took Alex to finish the race.
Let B = the time it took Belinda to finish the race.

$B = 48 + 5$

$B = 53$

It took Belinda 53 seconds to finish the race.

$A = 53 - 10$

$A = 43$

It took Alex 43 seconds to finish the race.
Write/Explain: I used the Work Backward strategy to find how long it took Alex to finish the race. I found that it took Belinda 53 seconds and Alex 43 seconds to finish the race. I found the numbers that made these sentences true. It was 1:00 p.m. and 53 seconds when Belinda finished the race. It was 1:00 p.m. and 43 seconds when Alex finished the race.

Score the Answer.

According to the rubric, from **1** to **3**, what score would you give Miguel? Explain why you gave that score.

____

____

____

____

____

Make it a 4! Rewrite.

____

____

____

____

____

____

Use the rubric on *page 13* to score this work.

3. Ashwin Tries It.

Remember, there is often more than one way to solve a problem. Here is how Ashwin solved this problem.

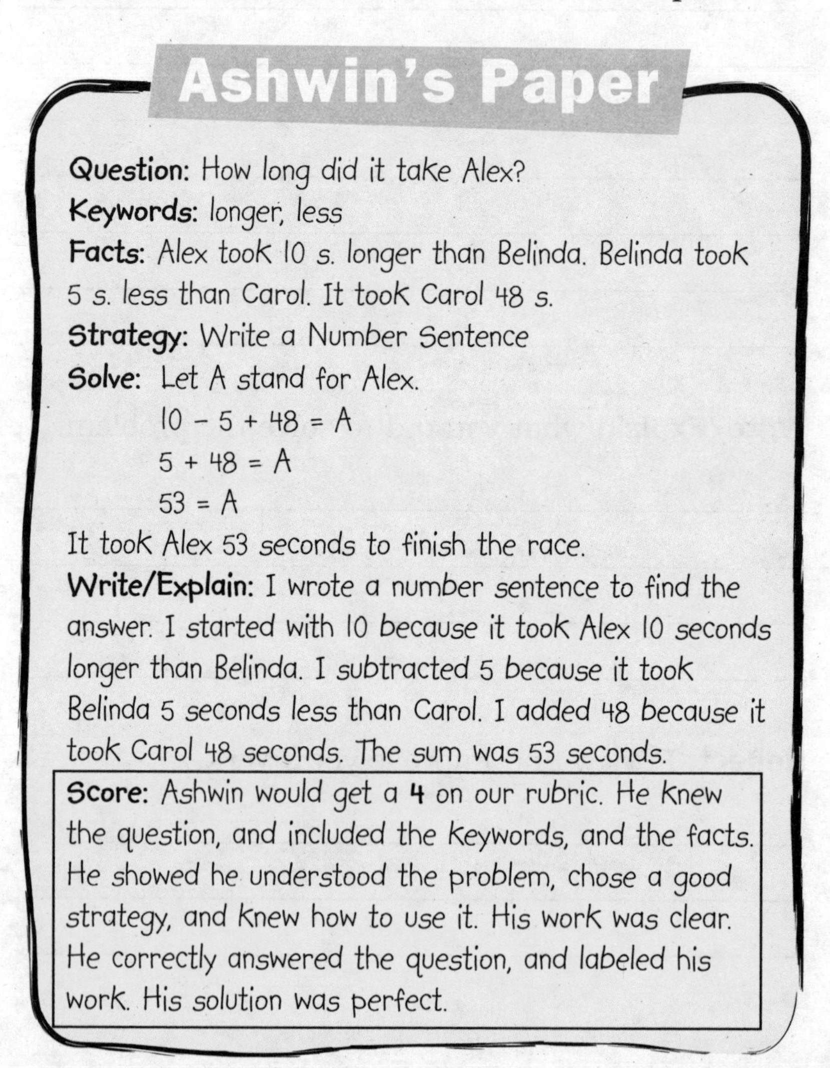

Ashwin's Paper

Question: How long did it take Alex?
Keywords: longer, less
Facts: Alex took 10 s. longer than Belinda. Belinda took 5 s. less than Carol. It took Carol 48 s.
Strategy: Write a Number Sentence
Solve: Let A stand for Alex.

$10 - 5 + 48 = A$
$5 + 48 = A$
$53 = A$

It took Alex 53 seconds to finish the race.
Write/Explain: I wrote a number sentence to find the answer. I started with 10 because it took Alex 10 seconds longer than Belinda. I subtracted 5 because it took Belinda 5 seconds less than Carol. I added 48 because it took Carol 48 seconds. The sum was 53 seconds.

Score: Ashwin would get a **4** on our rubric. He knew the question, and included the keywords, and the facts. He showed he understood the problem, chose a good strategy, and knew how to use it. His work was clear. He correctly answered the question, and labeled his work. His solution was perfect.

4. Answers to Parts 1 and 2.

Guided Problem #4

Alex and Belinda started the race at 1:00 p.m. It took Alex 10 seconds longer than Belinda to finish a race. Belinda took 5 seconds less than Carol to finish. Carol finished the race in 48 seconds. How long did it take Alex to finish the race?

Keywords: longer, less

1. Try It Yourself. (pages 72–73)

Question: How long did it take Alex?

Facts: It took Alex 10 s. longer than Belinda. It took Belinda 5 s. less than Carol.

It took Carol 48 s.

Strategy: Write Number Sentences

Solve: Belinda: 48 – 5 = 43

Alex: 43 + 10 = 53

It took Alex 53 seconds to finish the race.

Write/Explain: I know that it took Carol 48 seconds to finish the race. I can find how long it took Alex to finish the race by finding out how long it took Belinda. I used Number Sentences. Belinda took 5 seconds less than Carol: 48 – 5 = 43. It took Alex 10 seconds longer than Belinda: 43 + 10 = 53. It took Alex 53 seconds to finish the race.

2. Miguel Tries It. (page 74)

Score the Answer: Miguel would receive a **3** on the rubric. He included the keywords, the facts, the correct question, and labeled his work. He picked a good strategy, but used the wrong operations. He also did not realize that the time Alex and Belinda started was extra information which was not necessary in solving the problem.

Make it a 4! Rewrite.

B = 48 + 5 should have been B = 48 – 5.

A = B – 10 should have been A = B + 10.

So, B = 43. It took Belinda 43 seconds to finish the race.

A = 53. It took Alex 53 seconds to finish the race.

Guided Problem #5

Fred is playing with a number machine. He puts 1 number in and another comes out. This is what has happened so far.

Input	Output
3	12
5	20
7	28
9	36
12	?

What is the missing number? Explain your answer.

Keyword: ?

1. Try It Yourself.

Answer the questions below to get a score of **4**.

What **question** are you being asked?

____

What is the **keyword**?

____

What are the **facts** you need to solve the problem?

____

____

____

____

What **strategy** can you use to solve the problem?

____

____

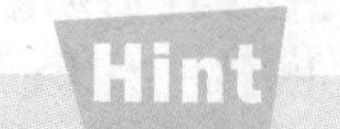

Possible answers include **Look for a Pattern**.

Solve the problem.

Write/Explain what you did to solve the problem.

Reflect. Review and improve your work.

2. Martha Tries It.

Score the Answer.

According to the rubric, from **1** to **3**, what score would you give Miguel? Explain why you gave that score.

Make it a 4! Rewrite.

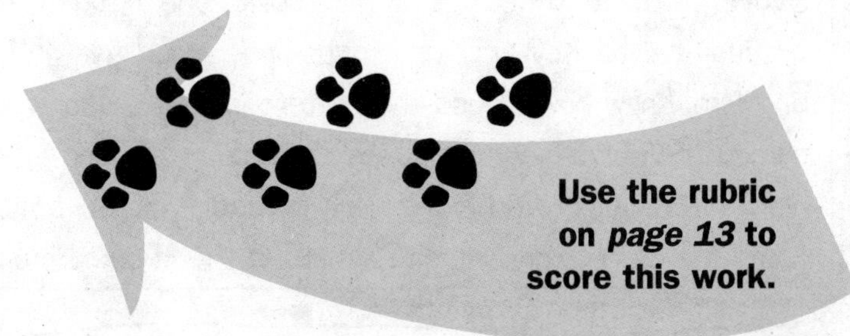

3. Patti Tries It.

Remember, there is often more than one way to solve a problem. Here is how Patti solved this problem.

Patti's Paper

Question: What is the Output when 12 is the Input?
Keyword: missing
Facts: 3 becomes 12, 5 becomes 20, 7 becomes 28, 9 becomes 36
Strategies: Find a Pattern, Write Number Sentences
Solve: The pattern is, the Output is 4 times greater than the Input. Subtract 12 – 9 = 3 and multiply 3 x 4 = 12. The difference in the output between 36 and the missing number is 12. Add 36 + 12 = 48. The missing number is 48.
Write/Explain: I Found a Pattern and Wrote Number Sentences. I found the difference between 12 and 9 in the Input column. I then multiplied the difference by 4 since the Output is always 4 times greater than the Input. Since 3 x 4 = 12, I added 12 + 36 = 48. 48 is the missing number.

Score: Patti would get a **4** on our rubric. She included the question asked, Keyword, and facts, understood the problem, knew how to find the pattern in the function table, named the strategies used, used them correctly, and labeled her work. She created extra work by adding steps, but still found the correct answer and explained the steps she used.

4. Answers to Parts 1 and 2.

Guided Problem #5

Fred is playing with a number machine. He puts 1 number in and another comes out. This is what has happened so far.

Input	Output
3	12
5	20
7	28
9	36
12	?

What is the missing number? Explain your answer.

Keyword: missing

1. Try It Yourself. (pages 76–77)

Question: When the Input is 12, what is the Output?

Fact: 9 becomes 36.

Strategy: Look for a Pattern

Solve: The number in the Output is 4 times greater than the Input. I can find the missing number by multiplying $12 \times 4 = 48$.

Write/Explain: I Found a Pattern in the table. For each Input, the Output is 4 times greater. To find the missing number I can multiply $12 \times 4 = 48$. I checked my answer and it is correct.

2. Martha Tries It. (page 77)

Score the Answer: Martha would get a **1**. She answered the question correctly, but she did not show any work. She did not give the keywords, the facts, or the question that was being asked. She did not show the strategy she used to find the answer. She did not explain how she found the answer.

Make it a 4! Rewrite.

Question: What is the Output when 12 is the Input?

Keyword: missing

Facts: 3 becomes 12, 5 becomes 20, 7 becomes 28, 9 becomes 36

I used Look for a Pattern strategy.

The pattern is, the Output is 4 times greater than the Input. Multiply $12 \times 4 = 48$ to find the missing number.

The missing number is 48.

Quiz Problems

Here are some problems for you to try. Keep your **rubric** handy while you solve the problem. Let's see if you can score a **4**.

1. Carlos has 18 more DVDs than Marc, and Robert has 6 fewer DVDs than Carlos. Marc has 36 DVDs. How many DVDs do Carlos and Robert have?

2. When you put a number into a number machine, another number comes out. Mabel is playing with the machine. Here are her results:

Input	Output
3	8
4	9
5	10
6	?
8	?
10	?

(a) Find the numbers that would have come out of the machine when 6, 8 and 10 were put in.

(b) Write a Number Sentence that tells what the machine's rule is.

(c) Write a word sentence that tells what the rule is.

3. Here is the puzzle that Dave gave to Suzy. "I am thinking of 2 numbers. Their product is 36 and their sum is 13. What are the 2 numbers?"

4. Sam has 48 bouncy balls. He has twice as many red ones as he has yellow ones. How many of each color does he have?

5. Ling is making a necklace using circles, triangles, and squares. Ling is using twice as many triangles as circles. She is using 3 times as many squares as circles. She used 18 shapes in all. How many of each shape did Ling use?

6. The table shows how much money Manny received on each of his birthdays.

Birthday	Money Received
1	$1
2	$2
3	$4
4	$8

If the pattern continues, how much money will Manny receive on his eighth birthday?

7. Tad is designing a border for his wall. He has a rectangle, a circle, a triangle, and a pentagon. The design follows in that order. Which is the 17th shape in the design?

7. Geometry

Look around your home, your school, your neighborhood. What **shapes** do you see? Squares? Cylinders? Triangles? What **lines** do you see? Some that never meet? Some that intersect? Some shorter? Some longer? What **angles** do you see? Some that are right angles? Some that are not? You are looking at geometry. Geometry is part of mathematics. Geometry is part of your everyday life.

Here is a geometry problem that might be on your tests. Let's see how we can solve this problem and get a perfect score of **4**.

We will use our **rubric** to **double-check**.

Modeled Problem

Miguel drew a **square**. Then he drew a **line segment** from the lower-left **corner** to the upper-right **corner** of the square. What 2 **shapes** did the **line segment** create? Identify the **shapes** as **specifically** as you can.

Keywords: square, line segment, corner, shapes, specifically

1. Read and Think

What **question** were we asked?

What are the **keywords**?

- **square, line segment, corner, shapes, specifically**

What 2 shapes were created by a line segment in a square?

What **facts** were we given?

- **We know that 1 line segment was drawn inside a square from the left lower corner to the right upper corner.**

2. Select a Strategy

We can **Draw a Picture**.

3. Solve

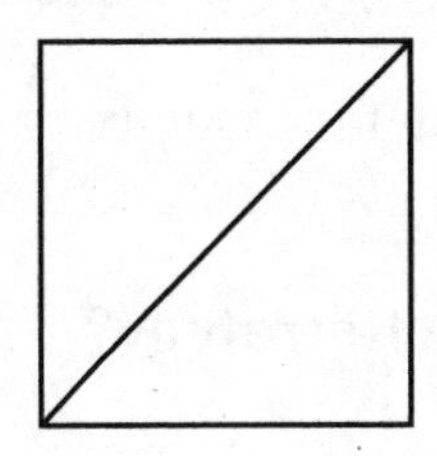

The 2 shapes are right triangles. We can measure the sides to see if the triangles are scalene or isosceles. Two of the sides of the triangle have the same length, so it is an isosceles triangle.

4. Write/Explain

We **Drew a Picture** of a square with a line segment from the lower left corner to the upper right corner. The shapes created are isosceles right triangles. Each triangle has 1 right angle. Two of the sides of each triangle have the same length.

5. Reflect

Let's review our work and answer.

- Did we show that we knew what the problem asked? **Yes.**
- Did we know what the keywords were? **Yes.**
- Did we show that we knew what facts were given? **Yes.**
- Did we name and use the correct strategy? **Yes.**
- Was our mathematics correct? **Yes. We checked it. It was correct.**
- Did we label our work? **Yes.**
- Was our answer correct? **Yes.**
- Were all of our steps included? **Yes.**
- Did we write a good, clear explanation of our work? **Yes.**

Score

This would earn a perfect **4** on our rubric. All the work is correct. The strategy was a good one. Miguel did not leave out any steps.

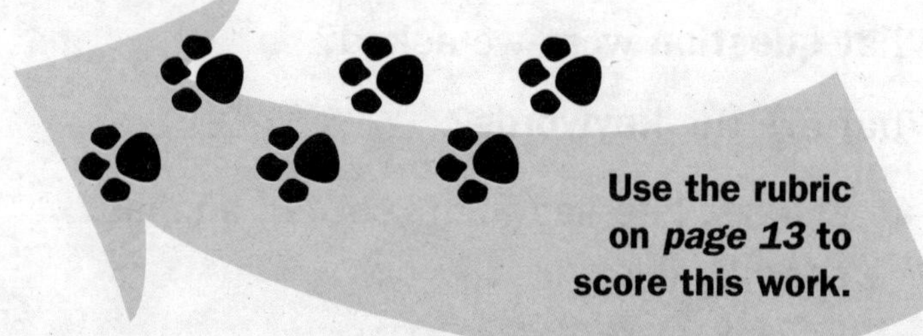

Here are some **Guided Open-Ended Math Problems**.

For each problem there are **four parts**. In the **first part**, you will solve the problem with guided help. In the **second part**, you will score and correct a solution with guided help. The **third part** shows one solution that scores a perfect **4**. This solution may or may not differ from your way. The **fourth part** has *answers* to the **first** and **second parts** so you can check your work.

Guided Problem #1

Jan is going to make a shape on a coordinate grid. She was told that the figure had 4 right angles. She wrote ordered pairs at (1, 2), (3, 2), (1, 4), and (3, 4). She connected the pairs in the order in which she wrote them. What shape did she make? Answer as specifically as you can.

1. Try It Yourself.

Answer the questions below to get a score of **4**.

What **question** are you being asked?

____

What are the **keywords**?

____

____

____

What are the **facts** you need to solve the problem?

____

____

____

____

What **strategy** can you use to solve the problem?

____

____

Hint

Possible answers include: **Draw a Picture**, **Logical Thinking**.

Solve the problem.

Write/Explain what you did to solve the problem.

Reflect. Review and improve your work.

2. Michelle Tries It.

Michelle's Paper

Question: What shape can be made from the ordered pairs?
Keywords: coordinate grid, ordered pairs, specifically
Facts: There are ordered pairs at (1, 2), (3, 2), (1, 4), (3, 4).
Strategy: Draw a Picture

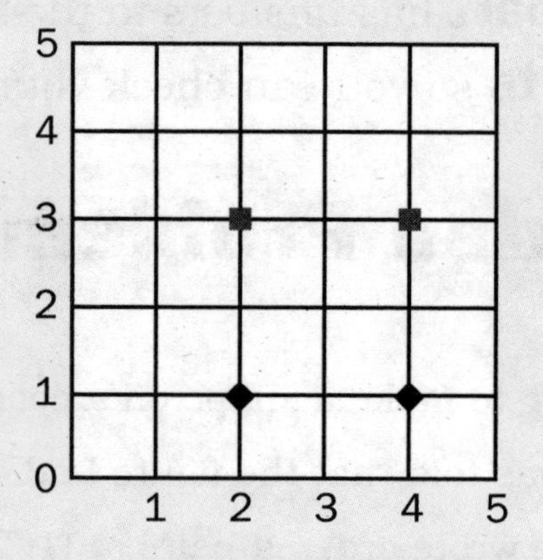

Write/Explain: I used a Drawing Strategy. I plotted the ordered pairs on a coordinate grid. Then I connected the points. The figure that I created had 4 equal sides, so it is a square.

Score the Answer.

According to the rubric, from **1** to **3**, what score would you give Michelle? Explain why you gave that score.

Make it a 4! Rewrite.

3. Larry Tries It.

Remember, there is often more than one way to solve a problem. Here is how Larry solved this problem.

Larry's Paper

Question: What shape can be made from the ordered pairs with 4 right angles?

Keywords: coordinate grid, ordered pairs, specifically

Facts: There are ordered pairs at (1, 2), (3, 2), (1, 4), (3, 4).

Strategy: I used logical thinking.

Solve: There are 4 ordered pairs. A figure with 4 sides is a quadrilateral.
The difference between the first coordinates is 2.
The difference between the second coordinates is 2.
All sides are equal, and there are 4 right angles, so the figure is a square.

Write/Explain: I used Logical Thinking. I counted the number of points and determined the figure was a quadrilateral. I subtracted the difference between the pairs of coordinates. The differences were both 2, so all sides are equal. A quadrilateral with 4 equal sides is a square.

Score: Larry's solution would earn a **4** on a test. He identified the question that was asked, the keywords, and the facts. He picked a good strategy, and explained how he used it. He clearly explained the steps taken to solve the problem. Larry labeled his work. It is perfect!

4. Answers to Parts 1 and 2.

Guided Problem #1

Jan is going to make a shape on a coordinate grid. She was told that the figure had 4 right angles. She wrote ordered pairs at (1, 2), (3, 2), (1, 4), and (3, 4). She connected the pairs in the order in which she wrote them. What shape did she make? Answer as specifically as you can.

Keywords: coordinate grid, ordered pairs, specifically

1. Try It Yourself. (pages 85–86)

Question: What shape can be made?

Facts: Points are at (1, 2), (3, 2), (1, 4), and (3, 4).

Strategy: Draw a Picture

Solve: When I connect the ordered pairs, I find that each point is exactly the same distance from the other points. The figure has 4 sides, so it is a square.

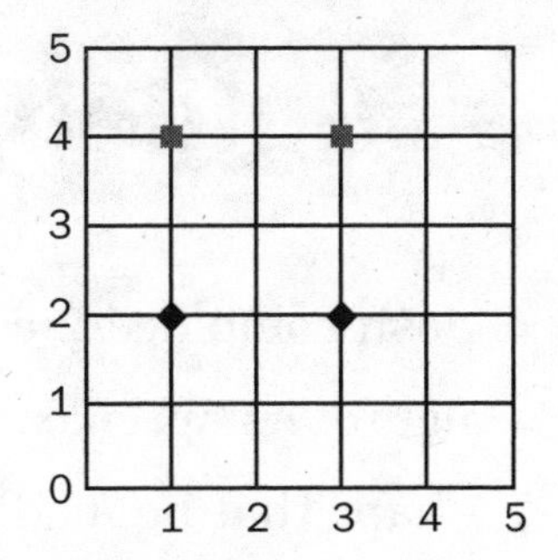

Write/Explain: I Drew a Picture. I plotted the ordered pairs on a coordinate grid. Then I connected the points. The figure that I created had 4 equal sides, which is a square.

2. Michelle Tries It. (pages 86–87)

Score the Answer: I would give Michelle a **3**. She gave the keywords, listed the facts, and knew what question was being asked. Her answer was correct and she labeled her work. However, she plotted the ordered pairs incorrectly. She confused the first and second coordinates with each ordered pair.

Make it a 4! Rewrite.

Everything other than the ordered pairs was correct. She needs to plot the correct ordered pairs.

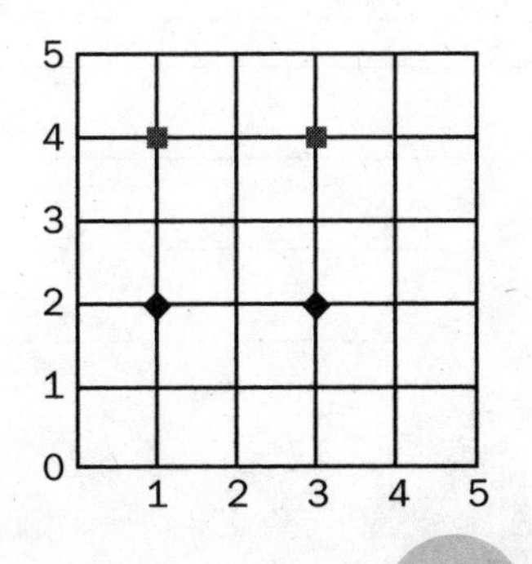

Guided Problem #2

Josh combined 2 congruent squares to form 1 figure. As specifically as possible, identify the shape that he made.

Keywords: 

1. Try It Yourself.

Answer the questions below to get a score of **4**.

What **question** are you being asked?

What are the **keywords**?

What are the **facts** you need to solve the problem?

What **strategy** can you use to solve the problem?

Hint

Possible answers include: **Draw a Picture, Logical Thinking**.

Solve the problem.

Write/Explain what you did to solve the problem.

Reflect. Review and improve your work.

2. Angela Tries It.

Angela's Paper

Question: What shape is formed by 2 congruent squares?

Keywords: combined, congruent, squares, form, specifically

Facts: 2 congruent squares are formed.

Strategy: Draw a Picture

Solve: I Drew a Picture of two congruent squares.

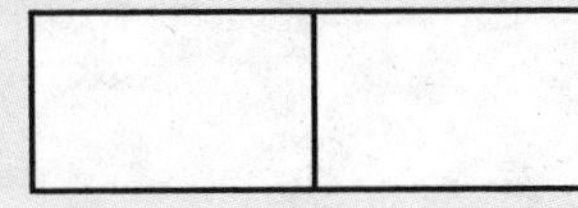

Write/Explain: I Drew a Picture of 2 congruent squares and combined them. Congruent means they are the same shape, and a rectangle has 4 sides and 4 right angles. The shape that is formed is a rectangle.

Score the Answer.

According to the rubric, from **1** to **3**, what score would you give Angela? Explain why you gave that score.

____

____

____

____

Make it a 4! Rewrite.

____

____

____

____

____

____

____

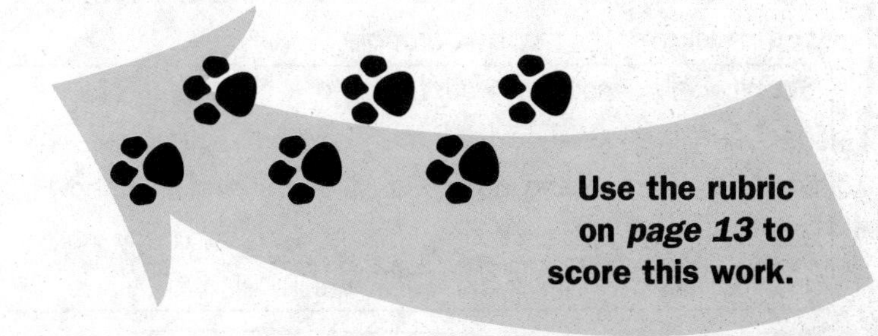

3. Felix Tries It.

Remember, there is often more than one way to solve a problem. Here is how Felix solved this problem.

Felix's Paper

Question: What shape is made from two congruent squares.
Keywords: combined, congruent, squares, form, specifically
Facts: Two congruent squares are combined.
Strategy: I used Logical Thinking.
Solve: A square has 4 equal sides. Congruent means the same size and shape. If 2 squares are combined the length will be twice as long as the width. I can plot a square on a grid at (1, 1), (1, 2), (2, 1), and (2, 2). A second square can go at (2, 1), (2, 2), (3, 1), and (3, 2). So the new figure would have 4 sides with a length of 2 and a width of 1. The figure is a rectangle.
Write/Explain: I used Logical Thinking. I gave the ordered pairs for 2 congruent squares to show that the combined figure has 4 sides with a length twice as long as the width. That is a rectangle.

Score: Felix's solution would earn a **4** on a test. He identified the question that was asked, the keywords, and the facts, and picked a correct strategy, explaining how it was used. He gave an example of how his strategy works and explained what he did. Felix also labeled his work.

4. Answers to Parts 1 and 2.

Guided Problem #2

Josh combined 2 congruent squares to form 1 figure. As specifically as possible, identify the shape that he made.

Keywords: combined, congruent, squares, form, specifically

1. Try It Yourself. (page 90)

Question: What shape is made from 2 congruent squares?

Facts: 2 congruent squares are combined to make one figure

Strategy: Draw a Picture

Solve:

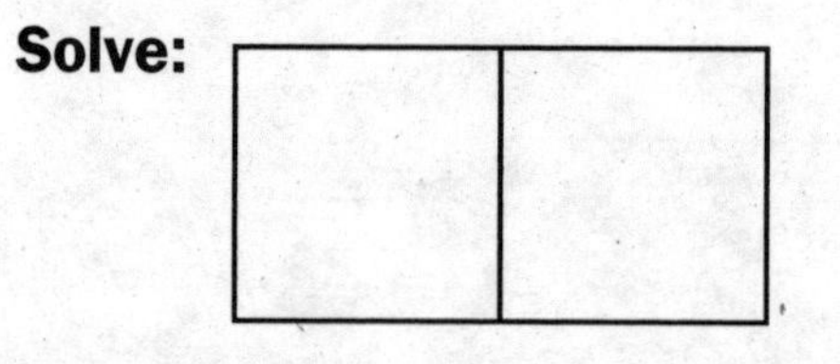

The new figure is a rectangle.

Write/Explain: I Drew a Picture of 2 congruent squares. I combined the squares, and the new shape had a length that was longer than the width. Since the new shape has 4 sides, all with right angles, it is a rectangle.

2. Angela Tries It. (page 91)

Score the Answer: I would give Angela a **3**. She gave the keywords, listed the facts, and knew what question was being asked. Her answer was correct. However, she drew 2 congruent rectangles, not squares.

Make it a 4! Rewrite.

Everything other than the drawing was correct. She needs to combine 2 congruent squares.

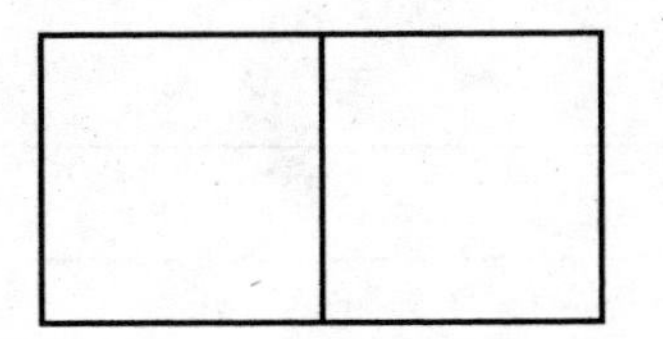

Guided Problem #3

Mrs. Davidson described to her students the following mystery figure.

"The figure I am thinking of is a 3-dimensional figure. It has 1 curved side and 2 flat sides. What figure am I thinking of?" Explain your answer.

Keywords:

1. Try It Yourself.

Answer the questions below to get a score of **4**.

What **question** are you being asked?

___

What are the **keywords**?

___

___

___

What are the **facts** you need to solve the problem?

____

What **strategy** can you use to solve the problem?

____

____

Hint

Possible answers include: **Make an Organized List, Draw a Picture,** and **Use Logical Thinking**.

Solve the problem.

____

____

____

____

Write/Explain what you did to solve the problem.

____

____

____

Reflect. Review and improve your work.

____

____

2. Mike Tries It.

Mike's Paper

Score the Answer.

According to the rubric, from **1** to **3**, what score would you give Mike? Explain why you gave that score.

____

____

____

____

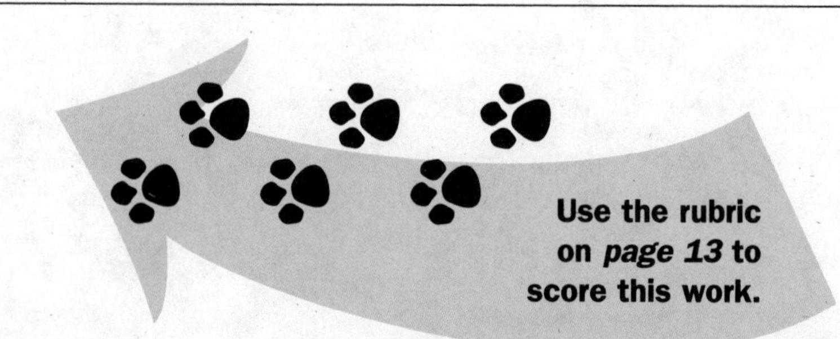

Make it a 4! Rewrite.

3. Tyrone Tries It.

Remember, there is often more than one way to solve a problem. Here is how Tyrone solved this problem.

Tyrone's Paper

Question: What 3-dimensional figure has 1 curved side and 2 flat faces?

Keywords: 3-dimensional figure, curved side, flat face

Facts: A 3-dimensional figure has 1 curved side and 2 flat faces.

Strategy: Logical Thinking

Solve: The only 3-dimensional figures with flat faces and curved sides are cylinders and cones. A can has 2 flat faces and 1 curved side. A can is a cylinder.

Write/Explain: I used Logical Thinking. I gave a real-world example of a 3-dimensional figure with 2 flat faces and 1 curved side. Then I identified the 3-dimensional figure.

Score: Tyrone would earn a **4** on our rubric. He identified the question that was asked, the Keywords, and the facts, picked a good strategy, explained how he used it, and labeled his work. He explained how he determined which shape was described by Mrs. Davidson.

4. Answers to Parts 1 and 2.

Guided Problem #3

Mrs. Davidson described to her students the following mystery figure.

"The figure I am thinking of is a 3-dimensional figure. It has 1 curved side and 2 flat sides. What figure am I thinking of?" Explain your answer.

Keywords: 3-dimensional figure, curved, flat faces

1. Try It Yourself. (pages 93–94)

Question: What 3-dimensional figure has 1 curved side and 2 flat faces?

Facts: A 3-dimensional figure has 1 curved side and 2 flat faces.

Strategy: Make an Organized List.

Solve:

Figure	Curved Sides	Flat Sides
Cube	No	Yes
Rectangular Prism	No	Yes
Sphere	Yes	No
Cylinder	Yes	Yes
Cone	Yes	Yes
Pyramid	No	Yes

The figures with curved sides and flat faces are cylinders and cones. A cylinder has 2 flat faces, so that is the mystery figure.

Write/Explain: I Made an Organized List to show which 3-dimensional shapes have curved sides, flat faces, or both. I determined that a cylinder and a cone both have curved sides and flat faces. A cylinder has two flat faces: a top flat face and a bottom flat face. I chose it as the mystery figure.

2. Mike Tries It. (page 94)

Score the Answer: I would give Mike a **1**. He drew the correct figure, but he did not give any other information. He did not identify the cylinder. He did not explain how he figured out the figure was a cylinder.

Make it a 4! Rewrite.

Question: What 3-dimensional figure has 1 curved side and 2 flat faces?

Keywords: 3-dimensional figure, curved side, flat face

Facts: A 3-dimensional figure has 1 curved side and 2 flat faces.

Mike Drew a Picture of the figure that was described. The figure has flat faces on top and on bottom and it has a curved side in the middle. It is a cylinder.

Guided Problem #4

Washington Street and Adams Street never meet and remain the same distance apart. This means they are parallel.

Jefferson Avenue crosses both streets at a right angle. This means they are perpendicular.

Jefferson Avenue and Madison Avenue never meet and remain the same distance apart. Madison Avenue crosses both Washington Street and Adams Street.

Which streets form parallel lines? Which streets form perpendicular lines?

Keywords: ? ?

Turn the page to begin the exercises.

7. Geometry

1. Try It Yourself.

Answer the questions below to get a score of **4**.

What **questions** are you being asked?

What are the **keywords**?

What are the **facts** you need to solve the problem?

What **strategy** can you use to solve the problem?

Hint

Possible answers include **Draw a Picture, Make an Organized List,** and **Use Logical Thinking**.

Solve the problem.

Write/Explain what you did to solve the problem.

Reflect. Review and improve your work.

2. Ellen Tries It.

Ellen's Paper

Questions: Which streets are parallel? Which streets are perpendicular?
Keywords: crosses, right angle, form, parallel lines, perpendicular lines
Facts: Washington St. and Adams St. never meet and remain the same distance apart.
Jefferson Ave. crosses both streets at a right angle.
Jefferson Ave. and Madison Ave. never meet and remain the same distance apart.
Madison Ave. crosses both Washington St. and Adams St.
Strategy: I Made an Organized List.
Solve:

Street 1	Street 2	Relationship	Type of Lines Formed
Washington	Adams	Never meet	Parallel
Washington	Jefferson	Meet at right angle	Perpendicular
Jefferson	Madison	Never meet	Parallel
Washington	Madison	Meet at right angle	Perpendicular

Write/Explain: I Made an Organized List to show the lines that can be formed by the streets. I know that lines that never meet and remain the same distance apart are parallel. Lines that meet at right angles are perpendicular. Washington is parallel to Adams, and Jefferson is parallel to Madison. Washington and Jefferson, and Washington and Madison are perpendicular.

Score the Answer.

According to the rubric, from **1** to **3**, what score would you give Ellen? Explain why you gave that score.

Make it a 4! Rewrite.

3. Walter Tries It.

Remember, there is often more than one way to solve a problem. Here is how Walter solved this problem.

Walter's Paper

Questions: Which streets are parallel to each other? Which streets are perpendicular to each other?
Keywords: crosses, right angle, form, parallel lines, perpendicular lines
Facts: Washington St. and Adams St. never meet and remain the same distance apart.
Jefferson Ave. crosses Washington St. and Adams St. at a right angle.
Jefferson Ave. and Madison Ave. never meet and remain the same distance apart.
Madison Ave. crosses both Washington St. and Adams St.
Strategy: Logical Thinking
Solve: The definition for parallel lines is lines that never meet and remain the same distance apart. Washington St. and Adams St. are parallel. Jefferson Ave. and Madison Ave. are parallel.
The definition for perpendicular lines is lines that cross at right angles. Jefferson Ave and Washington St. are perpendicular, as are and Jefferson Ave. and Adams St. Because Madison Ave. and Jefferson Ave. are parallel, Madison Ave. is also perpendicular to Washington St. and Adams St.

Write/Explain: I used Logical Thinking. I gave the definitions for parallel lines and perpendicular lines. I used the definitions to describe the streets that are parallel and the streets that are perpendicular.

Score: Walter would get a **4** on our rubric. He knew the questions that were asked and included the keywords and facts. Walter used an excellent strategy to find the correct answers. He explained how he used it. His explanation was detailed. He labeled his work.

4. Answers to Parts 1 and 2.

Guided Problem #4

Washington Street and Adams Street never meet and remain the same distance apart. This means they are parallel.

Jefferson Avenue crosses both streets at a right angle. This means they are perpendicular.

Jefferson Avenue and Madison Avenue never meet and remain the same distance apart. Madison Avenue crosses both Washington Street and Adams Street.

Which streets form parallel lines? Which streets form perpendicular lines?

Keywords: crosses, right angle, form, parallel lines, perpendicular lines

1. Try It Yourself. (page 98)

Questions: Which streets form parallel lines? Which streets form perpendicular lines?

Facts: Washington Street and Adams Street never meet and remain the same distance apart.

Jefferson Avenue crosses both streets at a right angle.

Jefferson Avenue and Madison Avenue never meet and remain the same distance apart. Madison Avenue crosses both Washington Street and Adams Street.

Strategy: Draw a Picture

Solve:

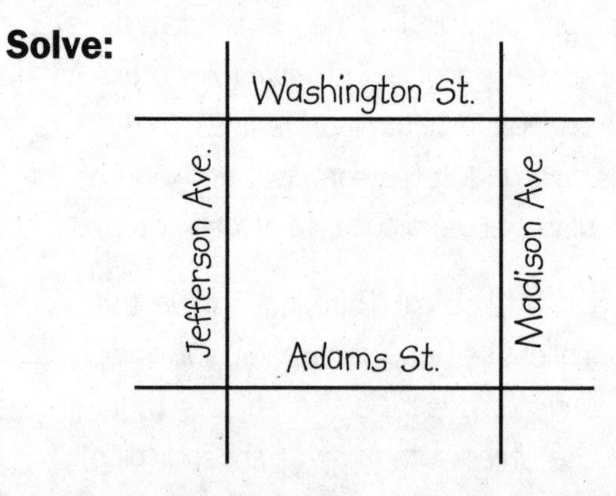

Parallel: Washington and Adams, and Jefferson and Madison

Perpendicular: Washington and Jefferson, Washington and Madison, Adams and Jefferson, Adams and Madison

Write/Explain: I Drew a Picture. I listed the street and avenue names as they were listed in the question. Then I determined that Washington and Adams were parallel as were Jefferson and Madison. Washington is perpendicular to Jefferson and Madison, and Adams is perpendicular to Jefferson and Madison.

2. **Ellen Tries It. (page 100)**

Score the Answer: I would give Ellen a **3**. She gave the questions that were asked, the keywords, the facts, and the strategy she used. Most of her answer was correct. She did not completely answer the question. She failed to explain the relationship between Adams St. and Jefferson Ave., and Adams St. and Madison Ave.

Make it a 4! Rewrite.

All of the information that Ellen gave is correct. Continue the table to show that Adams St. and Jefferson Ave. are perpendicular as are Adams St. and Madison Ave.

Quiz Problems

Here are some problems for you to try. Keep your **rubric** handy while you solve the problem. Let's see if you can score a **4**.

1. Leslie has drawn an equilateral triangle and a square. The sides of the triangle have the same length as the sides of the square. If Leslie combines the figures, what polygon does she create?

2. Marvin has drawn a rectangle that is 3 centimeters long and 2 centimeters wide. He has also drawn a rectangle that is 6 centimeters long and 4 centimeters wide. Are the rectangles congruent, similar but not congruent, or neither?

3. Tricia has plotted the ordered pair (0, 1) on a coordinate grid. She then plots the ordered pair (3, 4) on the grid. How can she draw a square?

4. Caitlin has a cube, sphere, cylinder, and cone. Explain which 3-dimensional figure will roll best.

5. Dennis has plotted an ordered pair at (1, 5). If he moves his ordered pair 3 units right and 2 units down, where does he plot his new ordered pair?

6. Liam has plotted the ordered pairs (1, 1), (4, 1), (2, 3), and (3, 3) on a coordinate grid. What polygon did he draw? Be as specific as possible.

7. How many squares can you find in this diagram? Hint: The answer is not 9.

8. Measurement

There are all kinds of measurements on tests. You have to know about **length,** which includes **inches, feet, yards, centimeters,** and **meters**. You have to know about **capacity,** such as **cups, pints, quarts, gallons,** and **liters**. You have to know about units like **ounces, pounds, grams,** and **kilograms** that are used to measure **weight** and **mass**. **Money, time,** and **temperature** are also kinds of measurement.

8. Measurement

Let's look at a problem in measurement that you might have to solve on a test. Let's see how to solve it. Remember, we want to get a perfect score of **4** on our scoring rubric.

Modeled Problem

Four girls ran in a race. When they finished, Janice was 20 feet **ahead** of Beth. Shelly was 25 feet **behind** Janice. Janice finished behind Tina. In what **order** did the 4 girls finish the race?

Keywords: ahead, behind, order

1. Read and Think

What **question** were we asked?

- **We were asked to find the order in which the girls finished the race.**

What are the **keywords**?

- **ahead**
- **behind**
- **order**

What **facts** were we given?

- **Janice finished 20 feet ahead of Beth.**
- **Shelly was 25 feet behind Janice. Janice finished behind Tina.**

2. Select a Strategy

The setting of the problem is a race, so let's use **Logical Thinking**. This is a good strategy to use because it will help to figure out the order in which the girls finish.

3. Solve

Let's use the initials of the girls' first names to indicate their places. The place to the left will be the person who finished first.

Shelly was 25 feet behind Janice, so Janice was ahead of Shelly.

J S

Janice was behind Tina.

T J S

Janice finished 20 feet ahead of Beth. We know that Shelly was 25 feet behind Janice, so Beth was 5 feet ahead of Shelly.

T J B S

The order from first to fourth was Tina, Janice, Beth, and Shelly.

4. Write/Explain

We used **Logical Thinking** to find out how the girls finished. We took each girl's position one step at a time to find where each placed. Janice finished behind Tina and ahead of Beth and Shelly. So Tina won the race and Janice was second. Beth was third and Shelly was fourth.

5. Reflect

Let's review our work and answer.

- Did we show that we knew what the problem asked? **Yes.**
- Did we know what the keywords were? **Yes.**
- Did we show that we knew what facts were given? **Yes.**
- Did we name and use the correct strategy? **Yes.**
- Was our mathematics correct? **Yes. We checked it. It was correct.**
- Did we label our work? **Yes.**
- Was our answer correct? **Yes.**
- Were all of our steps included? **Yes.**
- Did we write a good, clear explanation of our work? **Yes.**

Score

This solution would earn a **4** on our rubric. Logical Thinking is a good strategy to use in a race problem. It helped us to find the order in which the girls finished.

Here are some **Guided Open-Ended Math Problems**.

For each problem there are **four parts**. In the **first part**, you will solve the problem with guided help. In the **second part**, you will score and correct a solution with guided help. The **third part** shows one solution that scores a perfect **4**. This solution may or may not differ from your way. The **fourth part** has *answers* to the **first** and **second parts** so you can check your work.

Guided Problem #1

Mr. Ramirez walks 2,000 meters in the morning and 3 kilometers in the evening. Mrs. Ramirez walks 1,000 more meters each day than Mr. Ramirez. How many meters do they each walk each day?

Keywords: ? ?

Turn the page to begin the exercises.

Use the rubric on *page 13* to score this work.

8. Measurement

1. Try It Yourself.

Answer the questions below to get a score of **4**.

What **question** are you being asked?

What are the **keywords**?

What are the **facts** you need to solve the problem?

What **strategy** can you use to solve the problem?

Hint

Possible answers include **Divide and Conquer**, **Draw a Picture**.

Solve the problem.

Write/Explain what you did to solve the problem.

Reflect. Review and improve your work.

2. Lisa Tries It.

Lisa's Paper

Keywords: meters, kilometers, more

Facts: Mr. Ramirez walks 2,000 meters in the morning and 3 kilometers in the evening. Mrs. Ramirez walks 1,000 more meters than Mr. Ramirez.

Question: How many meters do they walk each day?

Solve: Mr. Ramirez walks 2,000 meters + 3 meters = 2,003 meters.

Mrs. Ramirez walks 2,003 meters + 1,000 meters = 3,003 meters.

They walk 2,003 meters + 3,003 meters = 5,006 meters

Write/Explain: I found how many meters Mr. Ramirez walks each day. I did this by adding 3 meters to 2,000 meters. Since Mrs. Ramirez walks 1,000 more meters than Mr. Ramirez, I added 1,000 meters + 2,003 meters = 3,003 meters. Then I added the total number of meters that they each had walked: 2,003 + 3,003 = 5,006.

Score the Answer.

According to the rubric, from **1** to **3**, what score would you give Lisa? Explain why you gave that score.

____

____

____

____

Make it a 4! Rewrite.

____

____

____

____

____

____

____

3. Travis Tries It.

Remember, there is often more than one way to solve a problem. Here is how Travis solved this problem.

Travis's Paper

Question: How many meters do they each walk a day?
Keywords: meters, kilometers, more
Facts: Mr. Ramirez walks 2,000 meters in the morning and 3 kilometers in the evening. Mrs. Ramirez walks 1,000 more meters than Mr. Ramirez.
Strategy: Draw a Picture
Solve: I Drew a Picture. I know that 1,000 meters = 1 kilometer, so I changed 3 kilometers to 3,000 meters.
I made a box for each 1,000 meters that Mr. Ramirez walks.

Mr. Ramirez				
M	M	E	E	E

Mr. Ramirez walks 5,000 meters each day.
Mrs. Ramirez walks 6,000 meters each day.

Write/Explain: I Drew a Picture. I converted meters to kilometers and made a box for each kilometer that Mr. Ramirez walks each day. I then added 1 box since Mrs. Ramirez walks 1,000 meters more than Mr. Ramirez. Then I converted kilometers back to meters.

Score: Travis's solution would earn a **4** on a test. He identified the question that was asked, the keywords, and the facts, picked a good strategy, explained how he used it, and labeled his work. He clearly explained the steps taken to solve the problem. It is perfect!

4. Answers to Parts 1 and 2.

Guided Problem #1

Mr. Ramirez walks 2,000 meters in the morning and 3 kilometers in the evening. Mrs. Ramirez walks 1,000 more meters each day than Mr. Ramirez. How many meters do they each walk each day?

Keywords: meters, kilometers, more

1. Try It Yourself. (pages 109–110)

Question: How many meters do they each walk each day?

Facts: Mr. Ramirez walks 2,000 meters in the morning.

Mr. Ramirez walks 3 km in the evening.

Mrs. Ramirez walks 1,000 more meters than Mr. Ramirez.

Strategy: Divide and Conquer

Solve: There are 1,000 meters in 1 kilometer, so 3 kilometers = 3,000 meters.

Mr. Ramirez walks 2,000 meters + 3,000 meters = 5,000 meters.

Mrs. Ramirez walks 5,000 meters + 1,000 meters = 6,000 meters.

Write/Explain: I used the Divide and Conquer strategy. First, I found how many meters Mr. Ramirez walks each day. I did this by converting kilometers to meters and then adding the 2 distances that he walks each day. Then I used that information to find how many meters Mrs. Ramirez walks each day.

2. Lisa Tries It. (page 111)

Score the Answer: I would give Lisa a **2**. She gave the keywords and listed the facts correctly. But she did not tell us the strategy she used. And, she did not convert kilometers to meters. This made her give the wrong answer. She did not answer the question that was asked.

Make it a 4! Rewrite.

Convert 3 kilometers to meters. There are 1,000 meters in 1 kilometer, so there are 3,000 meters in 3 kilometers. Mr. Ramirez walked 2,000 meters + 3,000 meters = 5,000 meters. Mrs Ramirez walked 5,000 meters + 1,000 meters = 6,000 meters.

Guided Problem #2

I walked 6 kilometers to get to the pool. The town pool is rectangular. It is 70 meters long and 30 meters wide. The pool manager decided to put a rectangular fence around the pool. The fence will be 100 meters long by 50 meters wide. How much fencing will he need?

Keywords: 

1. Try It Yourself.

Answer the questions below to get a score of **4**.

What **question** are you being asked?

___

What are the **keywords**?

___

___

What are the **facts** you need to solve the problem?

___

Hint

Possible answers include: **Draw a Picture** and **Write a Number Sentence**.

What **strategy** can you use to solve the problem?

___

___

Solve the problem.

___

___

___

Write/Explain what you did to solve the problem.

___

___

___

___

Reflect. Review and improve your work.

___

___

2. Dale Tries It.

Dale's Paper

Question: How many meters of fencing does the pool manager need?
Keywords: meters, rectangular
Facts: The pool is 70 meters by 30 meters.
6 kilometers to a house to a pool.
Strategy: Draw a Picture
Solve:

30 meters

70 meters

30 + 70 = 100.
The pool manager will need 100 meters of fencing.
Write/Explain: I Drew a Picture of the pool. I then added the length plus the width to find how much fencing is needed. Then I added the kilometers. The answer is 6 kilometers and 100 meters of fencing.

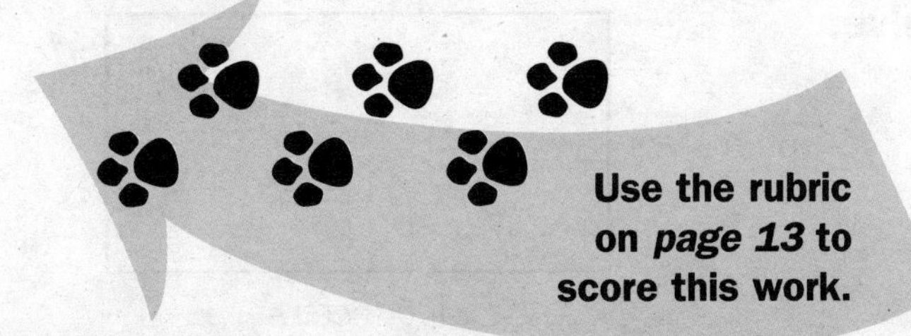

Use the rubric on *page 13* to score this work.

Score the Answer.

According to the rubric, from **1** to **3**, what score would you give Dale? Explain why you gave that score.

Make it a 4! Rewrite.

3. Tamika Tries It.

Remember, there is often more than one way to solve a problem. Here is how Tamika solved this problem.

Tamika's Paper

Question: What is the perimeter of the fence?
Keywords: meters, rectangular
Facts: The fence will be rectangular. It will be 100 meters by 50 meters.
Strategy: Write a Number Sentence
Solve: P = (100 meters x 2) + (50 meters x 2)
P = 200 meters + 100 meters = 300 meters
The pool manager needs 300 meters of fencing.
Write/Explain: I Wrote a Number Sentence to find the perimeter of the fence. I multiplied the length times 2, the width times 2, and then added the products to find the perimeter.

Score: Tamika's solution would earn a **4** on a test. She identified the question that was asked, the keywords, and the facts, picked a good strategy, explained how she used it, and labeled her work. She clearly explained the steps taken to solve the problem. It is perfect!

4. Answers to Parts 1 and 2.

Guided Problem #2

I walked 6 kilometers to get to the pool. The town pool is rectangular. It is 70 meters long and 30 meters wide. The pool manager decided to put a rectangular fence around the pool. The fence will be 100 meters long by 50 meters wide. How much fencing will he need?

Keywords: rectangular, meters

1. Try It Yourself. (page 114)

Question: What is the perimeter of the fence?

Facts: The fence is rectangular and is 100 meters by 50 meters.

Strategy: Draw a Picture

Solve:

50 meters

100 meters

I Drew a Picture.

50 meters + 50 meters + 100 meters + 100 meters = 300 meters

The pool manager needs 300 meters of fencing.

Write/Explain: I Drew a Picture of the fence. Since the fence is rectangular, I could figure out the missing sides. I added the 4 sides of the fence to find the perimeter. I made sure my arithmetic was correct and I gave the correct units in my answer.

2. Dale Tries It. (page 115)

Score the Answer: I would give Dale a **1**. He gave the wrong facts and did not answer the question correctly. Dale did not know that the perimeter is the measure of all 4 sides. He gave the dimensions of the pool, not the fence. Dale also didn't understand that the 6 kilometer walk to the pool was extra information not needed to answer the question.

Make it a 4! Rewrite.

The facts that are needed are that the fence will be 100 meters by 50 meters. The drawing should show a length of 100 meters and a width of 50 meters, not 70 and 30. To find the perimeter of the fence, add 100 meters + 100 meters + 50 meters + 50 meters = 300 meters. The pool manager needs 300 meters of fencing.

Guided Problem #3

Mrs. Bartlett made waffles for her family of 4. She had a container with 2 quarts of milk. She used 1 pint of milk to make the batter for the waffles. Everybody drank 1 cup of milk with their waffles. How many cups of milk were left after breakfast?

Keywords:

1. Try It Yourself.

Answer the questions below to get a score of **4**.

What **question** are you being asked?

____

____

What are the **keywords**?

What are the **facts** you need to solve the problem?

What **strategy** can you use to solve the problem?

Hint

Possible answers include: **Divide and Conquer, Draw a Picture,** and **Write a Number Sentence**.

Solve the problem.

Write/Explain what you did to solve the problem.

Reflect. Review and improve your work.

2. Laurie Tries It.

Laurie's Paper

Question: How much milk was left?
Keywords: quarts, pint, cup, left
Facts: There were 2 quarts of milk in a container. 1 pint of milk was used for the batter. Each of 4 people drank 1 cup of milk.
Strategy: Divide and Conquer
Solve: 2 quarts = 8 cups since 1 quart = 4 cups
1 pint + 4 cups = 5 cups
8 cups − 5 cups = 3 cups
There were 3 cups of milk left after breakfast.
Write/Explain: I used the Divide and Conquer strategy. I know that 2 quarts are equal to 8 cups. First, I found the number of cups of milk used by adding the amount used for the batter and the amount the family drank. Then I subtracted the number of cups of milk used from 8 cups.

Score the Answer.

According to the rubric, from **1** to **3**, what score would you give Laurie? Explain why you gave that score.

____

____

____

____

Make it a 4! Rewrite.

____

____

____

____

____

____

____

Use the rubric on *page 13* to score this work.

3. Pedro Tries It.

Remember, there is often more than one way to solve a problem. Here is how Pedro solved this problem.

Pedro's Paper

Question: How many cups of milk were left?

Keywords: quarts, pint, cup, left

Facts: A container had 2 quarts of milk. 1 pint was used to make the batter. Each of 4 family members drank 1 cup.

Strategy: I Drew a Picture.

Solve: Since 2 quarts = 8 cups, I made a rectangle with 8 parts.

Batter	Batter	Breakfast	Breakfast
Breakfast	Breakfast		

There are 2 cups of milk left.

Write/Explain: I Drew a Picture. I converted 2 quarts into 8 cups. I drew a rectangle with 8 equal parts to represent each of the 8 cups of milk in the container. I then filled in each box with a cup of milk that was used. I filled in 2 boxes with batter since 1 pint = 2 cups. There were 2 boxes left.

Score: Pedro would earn a **4** on our rubric. He identified the question that was asked, the keywords, and the facts, and picked a good strategy—Draw a Picture. He explained how his drawing worked, he gave the correct answer, and labeled his work.

4. Answers to Parts 1 and 2.

Guided Problem #3

Mrs. Bartlett made waffles for her family of 4. She had a container with 2 quarts of milk. She used 1 pint of milk to make the batter for the waffles. Everybody drank 1 cup of milk with their waffles. How many cups of milk were left after breakfast?

Keywords: quarts, pint, cup, left

1. Try It Yourself. (pages 117–118)

Question: How many cups of milk were left?

Facts: A container has 2 quarts of milk. 1 pint of milk was used for the batter.

4 people each drank 1 cup of milk.

Strategy: Divide and Conquer

Solve: A quart of milk contains 4 cups. So, 2×4 cups = 8 cups.

Amount used: 1 pint = 2 cups

Drinks: 4×1 cup = 4 cups

8 cups – (4 cups + 2 cups) = 2 cups

There were 2 cups of milk left after breakfast.

Write/Explain: I used the Divide and Conquer strategy. Two cups of milk were used for the waffles. The family drank 4 cups of milk at breakfast. All together, 6 cups of milk were used. A quart is equal to 4 cups, so 2 quarts are equal to 8 cups. I subtracted the amount used from the total and found that there were 2 cups left. I checked my arithmetic, which was correct.

2. Laurie Tries It. (page 119)

Score the Answer: I would give Laurie a **3**. She knew the keywords and the facts, answered the question that was asked, and labeled her work. She forgot to change 1 pint into 2 cups. Had she done that she would have gotten a **4**.

Make it a 4! Rewrite.

2 quarts = 8 cups since 1 quart = 4 cups

1 pint = cups

Start: 8 cups

Batter: 2 cups

Breakfast: 4 cups

Amount left: 8 cups – 6 cups = 2 cups

Guided Problem #4

The temperature was 78ºF at 10 a.m. At noon it was 84ºF. For the next 4 hours, the temperature rose 2 degrees each hour. What was the temperature at 3:00 p.m.?

Keywords:

1. Try It Yourself.

Answer the questions below to get a score of **4**.

What **question** are you being asked?

____

____

What are the **keywords**?

____

____

____

What are the **facts** you need to solve the problem?

____

____

What **strategy** can you use to solve the problem?

____

____

____

Hint

Possible answers include **Make a Table** and **Write a Number Sentence**.

Solve the problem.

Write/Explain what you did to solve the problem.

Reflect. Review and improve your work.

2. Carrie Tries It.

Carrie's Paper

Keywords: temperature, °F, rose

Facts: At 10 a.m., the temperature was 78°F. At noon, it was 84°F. For the next 4 hours the temperature rose 2 degrees each hour.

Strategy: I made a table.

Solve:

Time	Temperature
10 a.m.	78°F
12 noon	84°F
1 p.m.	86°F
2 p.m.	88°F
3 p.m.	90°F

92 − 78 = 14

The temperature rose 14°F.

Write/Explain: I made a table to show all of the information in the problem. I had to work out the temperature at each hour after 12 noon. I then subtracted the temperature at 10 a.m. from the temperature at 4 p.m., which was 4 hours after 12 noon.

Score the Answer.

According to the rubric, from **1** to **3**, what score would you give Carrie? Explain why you gave that score.

Make it a 4! Rewrite.

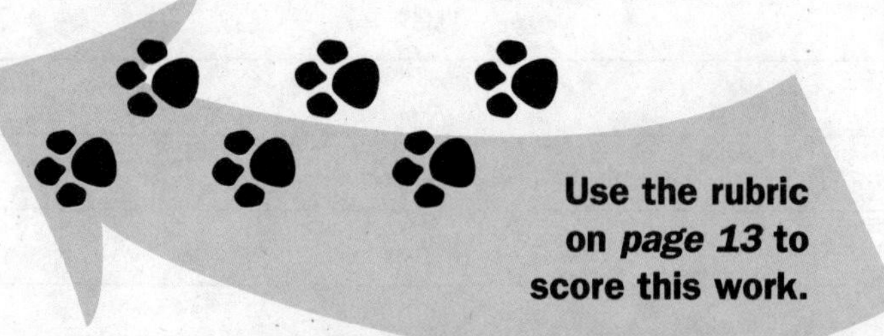

3. Courtney Tries It.

Remember, there is often more than one way to solve a problem. Here is how Courtney solved this problem.

Courtney's Paper

Question: What is the temperature at 3 p.m.?
Keywords: temperature, °F, rose
Facts: Facts: At noon, it was 84°F. For the next 4 hours the temperature rose 2 degrees each hour.
Strategy: Write a Number Sentence
Solve: $T = 84 + (3 \times 2)$
$T = 84 + 6 = 90$
The temperature at 3 p.m. was 90°F.
Write/Explain: I Wrote a Number Sentence. The temperature at noon was 84°F. The temperature rose 2° each hour for 4 hours. However, I only need 3 of those hours to solve the problem. I multiplied $2 \times 3 = 6$. Then I added the product to 84 to find the temperature at 3 p.m.

Score: Courtney would get a **4** on our rubric. She knew the question that was asked and included the keywords and the facts. She used an effective strategy to find the correct answer and labeled her work. Her explanation was detailed.

4. Answers to Parts 1 and 2.

Guided Problem #4

The temperature was 78°F at 10 a.m. At noon it was 84°F. For the next 4 hours, the temperature rose 2 degrees each hour. What was the temperature at 3:00 p.m.?

Keywords: temperature, °F, rose

1. Try It Yourself. (pages 122–123)

Question: What was the temperature at 3 p.m.?

Facts: It was 84°F at noon. The temperature rose 2 degrees each hour for 4 hours.

Strategy: Make a Table

Solve:

Time	Temperature
10 a.m.	78°F
12 noon	84°F
1 p.m.	86°F
2 p.m.	88°F
3 p.m.	90°F

At 3:00 p.m. the temperature was 90°F.

Write/Explain: I Made a Table to organize the data. I wrote the temperatures at 10 a.m. and 12 noon. Then I had to figure out the rest of the temperatures. I added 2 degrees every hour until I reached 3:00 p.m. I checked my arithmetic. It was right. I put the answer in degrees because that's what the problem asked for.

2. CarrieTries It. (pages 123–124)

Score the Answer: I would give Carrie a **3**. She gave the keywords, the strategy she used, and the facts. She labeled her work. She did not give the question and she did not answer the question that was asked. Her math was correct. She was supposed to find the temperature at 3 p.m. Even though she found this, as seen in the chart, she wrote down the wrong answer.

Make it a 4! Rewrite.

Question: What was the temperature at 3 p.m.?

Time	Temperature
10 a.m.	78°F
12 noon	84°F
1 p.m.	86°F
2 p.m.	88°F
3 p.m.	90°F

The temperature at 3 p.m. was 90°F.

Guided Problem #5

Mr. Wilson left his home at 10:45 a.m. to drive to a meeting. He arrived at the meeting at 12:15 p.m. How long was his trip?

Keywords: ? ?

1. Try It Yourself.

Answer the questions below to get a score of **4**.

What **question** are you being asked?

What are the **keywords**?

What are the **facts** you need to solve the problem?

____

____

What **strategy** can you use to solve the problem?

____

____

Hint

Possible answers include **Make a Table** and **Write a Number Sentence**.

Solve the problem.

____

____

____

Write/Explain what you did to solve the problem.

____

____

____

Reflect. Review and improve your work.

____

____

2. Jimmy Tries It.

Jimmy's Paper

Question: How long was his trip?

Keywords: left, arrived

Facts: Mr. Wilson left at 10:45 p.m. He arrived at 12:15 p.m.

Strategy: Write a Number Sentence

Solve: 12:15 – 10:45. I have to subtract the 2 times.

```
   12:15
 - 10:45
 -------
    1:70
```

1:70 = 2 hours and 50 minutes

Write/Explain: I Wrote a Number Sentence and subtracted 10:45 from 12:15 and got 1:70. 1 hour and 70 minutes is equal to 2 hours and 50 minutes.

Use the rubric on *page 13* to score this work.

Score the Answer.

According to the rubric, from **1** to **3**, what score would you give Jimmy? Explain why you gave that score.

Make it a 4! Rewrite.

3. Tracey Tries It.

Remember, there is often more than one way to solve a problem. Here is how Tracey solved this problem.

Tracey's Paper

Question: How long was his trip?

Keywords: left, arrived

Facts: Mr. Wilson left at 10:45 a.m. He arrived at 12:15 p.m.

Strategies: Make a List, Work Backward

Solve: 12:15 p.m. to 12 noon is 15 minutes
12:00 noon to 11 a.m. is 1 hour
11 a.m. to 10:45 a.m. is 15 minutes
15 minutes + 1 hour + 15 minutes = 1 hour and 30 minutes

Write/Explain: I made a list and worked backward. I counted back the minutes to 12 noon. I counted the hour back to 11 a.m. I counted back the minutes to 10:45 a.m. I added the times to find how long Mr. Wilson's trip was.

Score: Tracey would get a **4** on our rubric. She included the keywords, and the facts, and knew the question that was asked. She used effective strategies to find the correct answer. Her explanation was detailed. She labeled her work.

4. Answers to Parts 1 and 2.

Guided Problem #5

Mr. Wilson left his home at 10:45 a.m. to drive to a meeting. He arrived at the meeting at 12:15 p.m. How long was his trip?

Keywords: left, arrived

1. Try It Yourself. (pages 126–127)

Question: How long was his trip?

Facts: Mr. Wilson left at 10:45 a.m. He arrived at 12:15 p.m.

Strategy: I Made a Table.

Solve:

10:45 a.m. to 11:00 a.m.	15 minutes
11:00 a.m. to 12 noon	1 hour
12 noon to 12:15 p.m.	15 minutes

15 minutes + 1 hour + 15 minutes = 1 hour and 30 minutes

The trip took Mr. Wilson 1 hour and 30 minutes.

Write/Explain: I Made a Table and started at 10:45. I found the elapsed time until the next hour. Next, I counted the hours until noon. Then I found the elapsed time from noon until the arrival time. I added the time as 1 hour and 30 minutes.

2. Jimmy Tries It. (pages 127–128)

Score the Answer: I would give Jimmy a **2**. He gave the keywords, the strategy he used, the facts, the question, and labeled his work. He did not know how to subtract time, which led to an incorrect answer.

Make it a 4! Rewrite.

To find how long Mr. Wilson had driven, subtract 12:15 from 10:45 by changing 12:15 to 11:75 and then subtracting.

```
  11:75
– 10:45
-------
   1:30
```

Mr. Wilson's trip lasted 90 minutes.

Quiz Problems

Here are some problems for you to try. Keep your **rubric** handy while you solve the problem. Let's see if you can score a **4**.

1. Ms. Williams has 3 pounds of meat to make hamburgers. If she uses 4 ounces of meat for each hamburger, how many hamburgers can she make?

2. Mrs. Alley wants to put up a triangular fence around her garden. Two sides of the triangle are 10 meters each. The third side is 5 meters. She has 30 meters of fencing available to use. Does she have enough fencing to do the job?

3. Maria is 4 feet 7 inches tall. Her brother Antonio is 9 inches taller than she. What is Antonio's height, in inches?

4. Doris had an appointment with her dentist Tuesday, June 26. Doris needed more work done, so she had to return a week after Thursday. On what date will she next see her dentist?

5. Lee bought a box of golf balls. Inside the box there were 3 small cartons, each containing 3 golf balls. If the weight of all of the golf balls is 1 pound 2 ounces, what is the weight of a single golf ball?

6. Frannie brought 2 quarts of fruit juice to a picnic. Steve brought 6 cups of fruit juice and Nathan brought 5 pints of fruit juice. Who brought the most fruit juice? Who brought the least fruit juice?

7. Sue started watching a movie at 3:25 p.m. The movie lasted 2 hours and 15 minutes. At what time did the movie end?

9. Data Analysis and Graphs

What would you do if you had a lot of **data** to analyze? How would you **make sense** of it? How would you make it worthwhile and not wasted? How would you organize it? The data should be organized in some way. You may choose to use **tables, line plots, bar graphs,** or **line graphs**. You should choose the way you think displays the data best. When you do so, you can study the data in a logical way, and use it to draw conclusions.

It is important that you be able to solve open-ended problems that deal with **data** and **graphs**. Let's look at a modeled problem and see how it is done.

Modeled Problem

Mrs. Browne made 4 desserts for dinner. She told Lauren to spin the spinner to decide which dessert the family would have tonight. Number 1 represents apple pie, 2 represents chocolate cake, 3 represents vanilla ice cream, and 4 represents rice pudding.

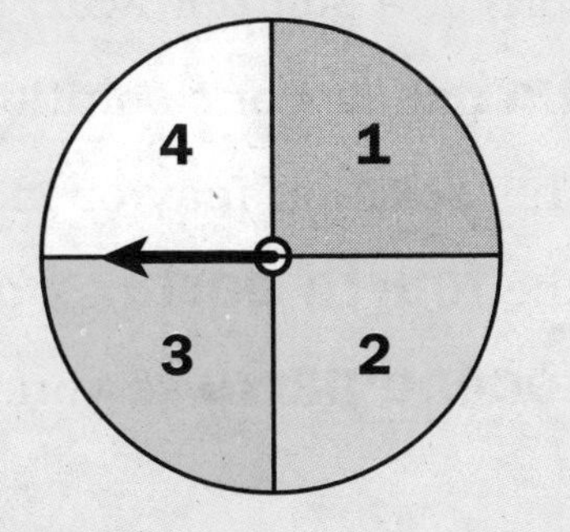

Which dessert is the family **most** likely to eat tonight?

Keyword: most

1. Read and Think

What **question** are we asked?

- **Which dessert is the family most likely to eat tonight?**

What is the **keyword**?

- **most**

What **facts** are we given?

- **There is a spinner with 4 equal parts:**
  1 stands for the apple pie,
  2 stands for the chocolate cake,
  3 stands for vanilla ice cream,
  4 stands for rice pudding.

2. Select a Strategy

We will use the **Draw a Picture** and some **Logical Thinking** to solve the problem.

3. Solve

The spinner is divided into 4 equal sections. Therefore each section is $\frac{1}{4}$ of the spinner. Since $\frac{1}{4} = \frac{1}{4}$, it is equally likely that the spinner will land on 1 of the numbers. It is equally likely that the Browne family will have any 1 of the desserts.

4. Write/Explain

We looked at the spinner. We used **Logical Thinking** to solve the problem. Each section is the same size. Each section is $\frac{1}{4}$ of the spinner. Therefore, it is equally likely that the Browne family will have any 1 of the desserts.

5. Reflect

Let's review our work and answer.

- Did we show that we knew what the problem asked for? **Yes.**
- Did we know what the keywords were? **Yes.**
- Did we show that we knew what facts were given? **Yes.**
- Did we name and use the correct strategy? **Yes.**
- Was our mathematics correct? **Yes. We checked it. It was correct.**
- Did we label our work? **Yes.**
- Was our answer correct? **Yes.**
- Were all of our steps included? **Yes.**
- Did we write a good, clear explanation of our work? **Yes.**

On the following page are **Guided Open-Ended Math Problems**.

For each problem there are **four parts**. In the **first part**, you will solve the problem with guided help. In the **second part**, you will score and correct a solution with guided help. The **third part** shows one solution that scores a perfect **4**. This solution may or may not differ from your way. The **fourth part** has *answers* to the **first** and **second parts** so you can check your work.

Guided Problem #1

In Emily's classroom the students voted on their favorite colors. The bar graph shows how they voted. How many more students voted for red than green?

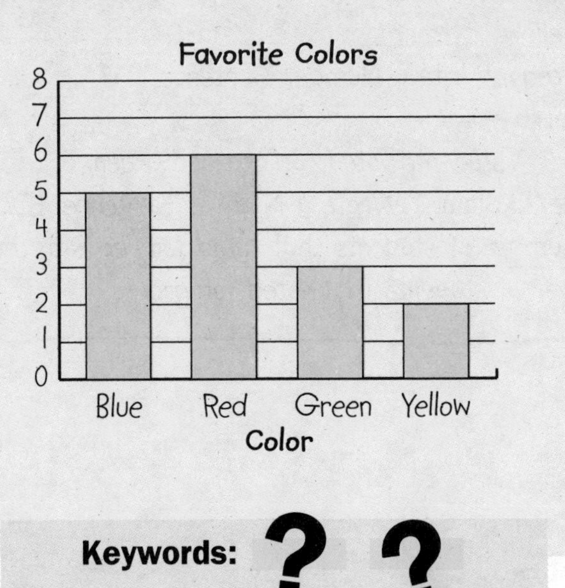

Keywords: **? ?**

1. Try It Yourself.

Answer the questions below to get a score of **4**.

What **question** are you being asked?

____

____

What are the **keywords**?

____

____

What are the **facts** you need to solve the problem?

____

____

____

____

9. Data Analysis and Graphs

What **strategy** can you use to solve the problem?

____

____

Hint

Possible answers include: **Write a Number Sentence**, **Draw a Picture**.

Solve the problem.

____

____

____

____

Write/Explain what you did to solve the problem.

____

____

____

Reflect. Review and improve your work.

____

____

____

____

2. Cate Tries It.

Cate's Paper

Question: How many students voted for red and green?

Keywords: more, graph

Facts: 5 students voted for blue, 6 students voted for red, 3 students voted for green, 2 students voted for yellow.

Strategy: Write a Number Sentence

Solve: 6 + 3 = 9

9 students voted for red and green.

Write/Explain: I Wrote a Number Sentence. I added the number of students that voted for red with the number of students that voted for green.

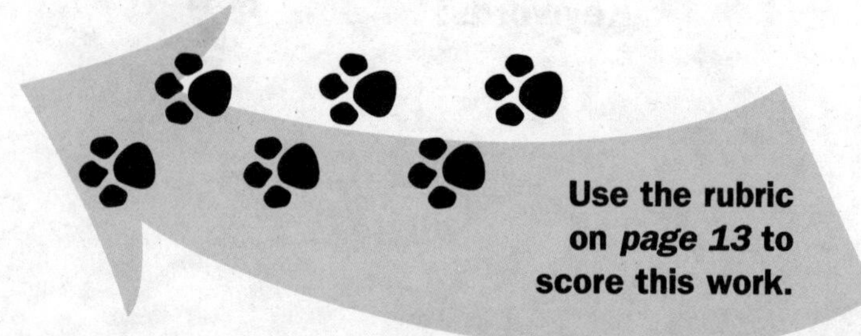

Score the Answer.

According to the rubric, from **1** to **3**, what score would you give Lisa? Explain why you gave that score.

____

____

____

____

Make it a 4! Rewrite.

____

____

____

____

____

____

____

3. Kathy Tries It.

Remember, there is often more than one way to solve a problem. Here is how Kathy solved this problem.

Kathy's Paper

Question: How many more students voted for red than for green?

Keywords: more, graph

Facts: 5 students voted for blue, 6 students voted for red, 3 students voted for green, and 2 students voted for yellow.

Strategy: I Drew a Picture.

Solve:

R	R	R
RG	RG	RG

3 more students voted for red than for green.

Write/Explain: I Drew a Picture. I drew 6 boxes to represent the number of students that voted for red. I put an R in each box. I then put a G for each vote that green got. The number of boxes that had an R only is the number of votes that red got that green did not.

Score: Kathy's solution would earn a **4** on a test. She identified the question that was asked, the keywords, and facts, and picked a strategy that worked. She explained how she used it. Although it would have been easier to just subtract 6 – 3, she clearly explained the steps taken to solve the problem. Finally, she labeled her work.

4. Answers to Parts 1 and 2.

Guided Problem #1

In Emily's classroom the students voted on their favorite colors. The bar graph shows how they voted. How many more students voted for red than green?

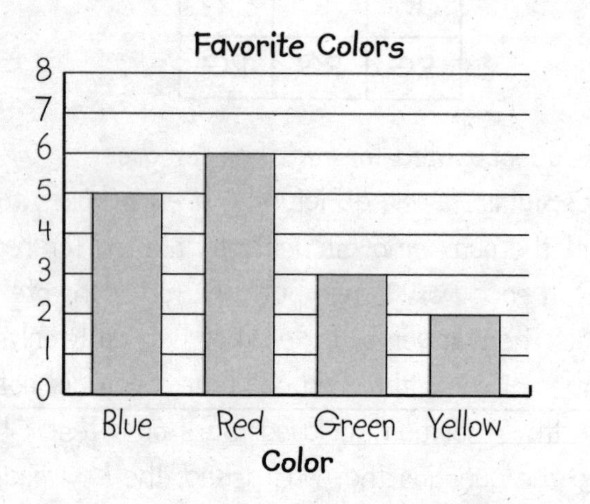

Keywords: more, graph

1. Try It Yourself. (pages 135–136)

Question: How many more students voted for red than green?

Facts: 5 students voted for blue. 6 students voted red.

3 students voted for green. 2 students voted for yellow.

Strategy: Write a Number Sentence

Solve: 6 students voted for red. 3 people voted for green.

$6 - 3 = 3.$

3 more students voted for red than voted for green.

Write/Explain: I Wrote a Number Sentence. I subtracted the number of students who voted for green from the number of students who voted for red.

2. Cate Tries It. (page 136)

Score the Answer: I would give Cate a **3**. She wrote the question to find, gave the keywords, and listed the facts correctly. She labeled her work. Her math was correct, but she did not answer the question that was asked. She was supposed to find how many more students voted for red than for green.

Make it a 4! Rewrite.

Question: How many more students voted for red than for green?

6 – 3 = 3.

Guided Problem #2

The table shows the heights of 4 students.

Heights of 4 Students

Student	Height (in Inches)
Larry	56
Mark	61
Nick	54
Omar	53

How much taller is Mark than Omar?

Keywords: 

1. Try It Yourself.

Answer the questions below to get a score of **4**.

What **question** are you being asked?

____

What are the **keywords**?

____

____

What are the **facts** you need to solve the problem?

____

____

What **strategy** can you use to solve the problem?

____

____

Hint

Possible answers include: **Write a Number Sentence** and **Act It Out**.

Solve the problem.

Write/Explain what you did to solve the problem.

Reflect. Review and improve your work.

2. Jill Tries It.

Jill's Paper

61- 53 = 8
Mark is 8 inches taller than Omar.

Score the Answer.

According to the rubric, from **1** to **3**, what score would you give Jill? Explain why you gave that score.

Make it a 4! Rewrite.

3. Emily Tries It.

Remember, there is often more than one way to solve a problem. Here is how Emily solved this problem.

Emily's Paper

Question: How much taller than Omar is Mark?

Keywords: heights, inches

Facts: Mark is 61 inches tall. Omar is 53 inches tall.

Strategy: Write a Number Sentence.

Solve: 53 + x = 61

53 + 8 = 61

Mark is 8 inches taller than Omar.

Write/Explain: I Wrote a Number Sentence to find how much taller Mark is than Omar. I wrote Omar's height and Mark's height as an addition sentence. Then I found the number that made the sentence true.

Score: Emily's solution would earn a **4** on a test. She identified the question that was asked, the keywords, and the facts, and picked a good strategy that worked. She explained how she used it. She labeled her work. Emily clearly explained the steps taken to solve the problem.

4. Answers to Parts 1 and 2.

Guided Problem #2

The table shows the heights of 4 students.

Heights of 4 Students

Student	Height (in Inches)
Larry	56
Mark	61
Nick	54
Omar	53

How much taller is Mark than Omar?

Keywords: heights, inches

1. Try It Yourself. (pages 139–140)

Question: How much taller is Mark than Omar?

Facts: Mark is 61 inches tall. Omar is 53 inches tall.

Strategy: Write a Number Sentence.

Solve: 61 – 53 = 8

Mark is 8 inches taller than Omar.

Write/Explain: I can Write a Number Sentence. I subtracted 61 – 53 = 8, so Mark is 8 inches taller than Omar.

2. Jill Tries It. **(page 140)**

Score the Answer: I would give Jill a **1**. She got the correct answer and labeled her work, but she did not give the keywords, facts, the question that was being asked, or a strategy. She did not write an explanation of her work.

Make it a 4! Rewrite.

Question: How much taller is Mark than Omar?

Keywords: heights, inches

Facts: Mark is 61 inches tall. Omar is 53 inches tall.

Write/Explain: I Wrote a Number Sentence. I subtracted Omar's height from Mark's height.

Guided Problem #3

Four students are collecting cans for the school recycling drive. The bar graph shows the number of cans that each student collected.

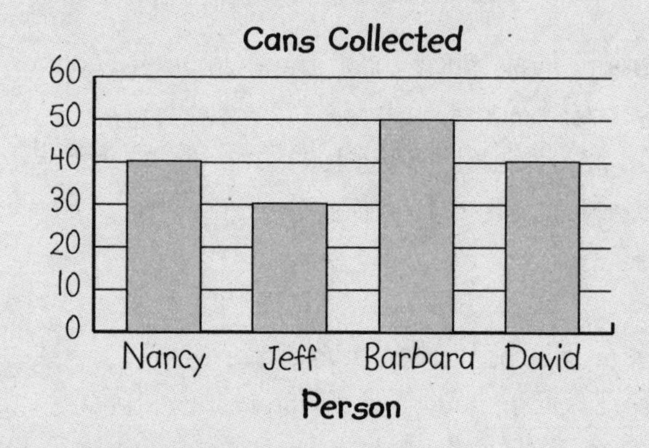

Who collected more cans: the boys or the girls? How many more?

Begin the exercises on the next page.

1. Try It Yourself.

Answer the questions below to get a score of **4**.

What **questions** are you being asked?

____

____

What are the **keywords**?

____

____

What are the **facts** you need to solve the problem?

____

____

____

____

What **strategy** can you use to solve the problem?

____

____

Hint

Possible answers include: **Divide and Conquer** and **Write a Number Sentence**.

Solve the problem.

____

____

____

____

____

Write/Explain what you did to solve the problem.

____

____

____

____

____

Reflect. Review and improve your work.

____

____

2. Bruce Solves It.

Bruce's Paper

Question: Who collected more cans: the boys or girls?

Question: How many more?

$40 + 30 = 70$

$50 + 40 = 90$

$90 - 70 = 20$

Keywords: more, bar graph

Facts: Nancy collected 40 cans. Jeff collected 30 cans. Barbara collected 50 cans. David collected 40 cans.

Strategy: I used the Divide and Conquer strategy.

Write/Explain: First I used number sentences to find out how many cans the boys and girls collected. I found the number of cans the boys collected. I found the number of cans the girls collected. Then I used another number sentence. I subtracted to find how many more cans were collected.

Use the rubric on *page 13* to score this work.

Score the Answer.

According to the rubric, from **1** to **3**, what score would you give Bruce? Explain why you gave that score.

____

____

____

____

____

Make it a 4! Rewrite.

____

____

____

____

____

____

____

3. Sandy Tries It.

Remember, there is often more than one way to solve a problem. See how Sandy solved this problem on the following pages.

Sandy's Paper

Questions: Who collected more cans: the boys or girls? How many more cans were collected by the group that collected more cans?

Keywords: more, bar graph

Facts: Nancy collected 40 cans.
Jeff collected 30 cans.
Barbara collected 50 cans.
David collected 40 cans.

Strategy: Write a Number Sentence.

Solve: 50 – 30 = 20

The girls collected 20 more cans than the boys.

Write/Explain: I used Logical Thinking and I Wrote a Number Sentence. Since Nancy and David collected the same number of cans, I did not include them in the answer. I compared the number of cans that Barbara and Jeff collected. Barbara collected 20 more cans than Jeff, so the girls collected 20 more cans than the boys.

Score: Sandy's solution would earn a **4** on a test. She identified the question that was asked, the keywords, and the facts, and picked strategies that worked. She explained how she used them. Sandy labeled her work. She clearly explained the steps taken to solve the problem.

4. Answers to Parts 1 and 2.

Guided Problem #3

Four students are collecting cans for the school recycling drive. The bar graph shows the number of cans that each student collected.

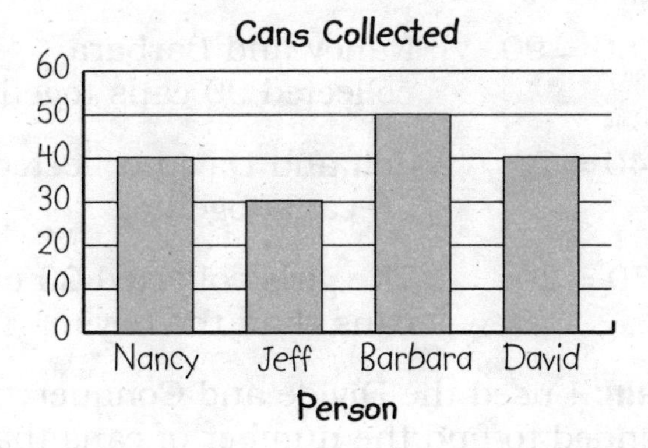

Who collected more cans: the boys or the girls? How many more?

Keywords: more, bar graph

9. Data Analysis and Graphs

1. Try It Yourself. (pages 142–143)

Questions: Did the boys or girls collect more cans?

How many more cans did one group collect than the other group?

Facts: Nancy collected 40 cans. Jeff collected 30 cans.

Barbara collected 50 cans. David collected 40 cans.

Strategy: Divide and Conquer

Solve: 40 + 50 = 90. Nancy and Barbara collected 90 cans together.

30 + 40 = 70. Jeff and David collected 70 cans together.

90 – 70 = 20. The girls collected 20 more cans than the boys.

Write/Explain: I used the Divide and Conquer strategy. I added to find the number of cans that the girls collected. I added to find the number of cans the boys collected. The girls collected more cans because 90 is greater than 70. I then subtracted 90 – 70 to find how many more cans the girls collected than the boys.

2. Bruce Tries It. (page 144)

Score the Answer: I would give Bruce a **3**. He gave the keywords and the facts, and answered the question that was asked. He used a good strategy and got the answer right. What he did wrong was he did not label his answers. He needed to say which group he was adding and what the numbers mean.

Make it a 4! Rewrite.

Keywords: more, bar graph

boys: 40 cans + 30 cans = 70 cans

girls: 50 cans + 40 cans = 90 cans

90 – 70 = 20

The girls collected 20 more cans than the boys.

Guided Problem #4

At the mall, Vivian bought a pair of black pants and a pair of tan pants. She also bought a striped shirt, a flowered shirt, and a polo shirt. Vivian said that for each of the next five days, she was going to wear a different outfit to school using the newly bought clothes. Can she really do that? Explain your answer.

Keywords: 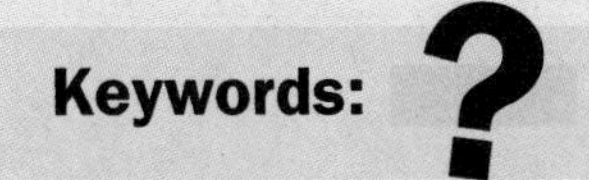

1. Try It Yourself.

Answer the questions below to get a score of **4**.

What **question** are you being asked?

____

What are the **keywords**?

____

What are the **facts** you need to solve the problem?

____

____

____

What **strategy** can you use to solve the problem?

____

____

Hint

Possible answers include **Make a List** and **Make a Table**.

Solve the problem.

____

____

Write/Explain what you did to solve the problem.

____

____

____

____

Reflect. Review and improve your work.

____

____

2. Jenna Tries It.

Jenna's Paper

Question: Can Vivian wear a different outfit each day for a week with the new clothes?
Keywords: different, five
Facts: Vivian has black pants and tan pants. She has a striped shirt, a flowered shirt, and a polo shirt. She wants to wear a different outfit each day for a week with the new clothes.
Strategy: Write a Number Sentence
Solve: She has 2 pairs of pants. She has 3 shirts.
$2 + 3 = 5.$
She can wear a different outfit each day for a week.
Write/Explain: I Wrote a Number Sentence: $2 + 3 = 5$. I added the number of pairs of pants and the number of shirts to get 5 outfits. Vivian can wear a different outfit to school for the next 5 days.

Use the rubric on *page 13* to score this work.

Score the Answer.

According to the rubric, from **1** to **3**, what score would you give Jenna? Explain why you gave that score.

Make it a 4! Rewrite.

3. Jamal Tries It.

Remember, there is often more than one way to solve a problem. Here is how Jamal solved this problem.

Jamal's Paper

Question: Can Vivian wear a different outfit each day for a week with the new clothes?

Keywords: different, five

Facts: Vivian has black pants and tan pants. She has a striped shirt, a flowered shirt, and a polo shirt. She wants to wear a different outfit each day for a week with the new clothes.

Strategy: Draw a Picture

Solve:

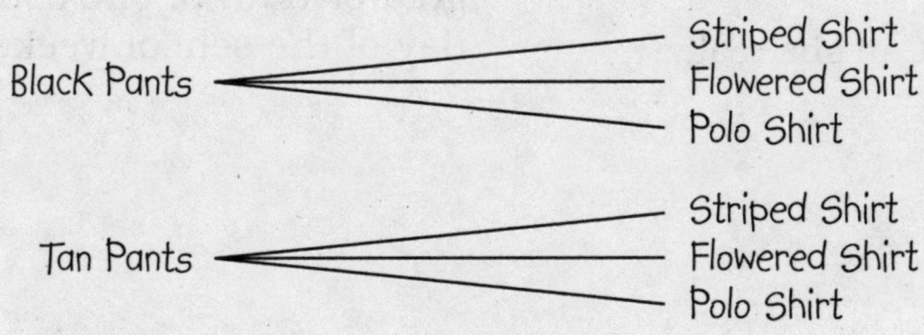

Black Pants — Striped Shirt
Black Pants — Flowered Shirt
Black Pants — Polo Shirt

Tan Pants — Striped Shirt
Tan Pants — Flowered Shirt
Tan Pants — Polo Shirt

The tree diagram shows that Vivian made six different outfits.

Write/Explain: I Drew a Picture, a tree diagram. The tree diagram showed the 6 different outfits that Vivian could make from her new clothes. Vivian has enough new clothes to wear a different new outfit each day for the next 5 days.

Score: Jamal would get a **4** on our rubric. He knew the question that was asked, and included the keywords and the facts. He used an effective strategy to find the correct answer. He explained how he used it. Jamal labeled his work. He clearly explained what he did. His work was perfect!

4. Answers to Parts 1 and 2.

Guided Problem #4

At the mall, Vivian bought a pair of black pants and a pair of tan pants. She also bought a striped shirt, a flowered shirt, and a polo shirt. Vivian said that for each of the next five days, she was going to wear a different outfit to school using the newly bought clothes. Can she really do that? Explain your answer.

Keywords: different, five

1. Try It Yourself. (page 147)

Question: Can Vivian wear a different outfit wearing the new clothes for 5 days?

Facts: Vivian has black pants and tan pants.

Vivian has a striped shirt, a flowered shirt, and a polo shirt.

She said she can wear a different outfit wearing the new clothes for 5 days.

Strategy: Make a List.

Solve:

black pants and striped shirt	tan pants and striped shirt
black pants and flowered shirt	tan pants and flowered shirt
black pants and polo shirt	tan pants and polo shirt

Write/Explain: I Made a List to find out how many different outfits Vivian could make. There were 6 different outfits that Vivian could make. Since 6 is greater than 5, she could wear a new outfit each day of the school week.

2. Jenna Tries It. (page 148)

Score the Answer: I would give Jenna a **3**. She gave the keywords, the facts, and the strategy she used, and she knew the question that was asked. She even got the answer right. She did not know how to find the number of outfits that Vivian could make.

Make it a 4! Rewrite.

Vivian could wear the following outfits:

black pants and striped shirt	tan pants and striped shirt
black pants and flowered shirt	tan pants and flowered shirt
black pants and polo shirt	tan pants and polo shirt

There are 6 outfits that Vivian could wear, so she could wear a different new outfit to school each day for the next 5 days.

Quiz Problems

Here are some problems for you to try. Keep your **rubric** handy while you solve the problem. Let's see if you can score a **4**.

1. Ted is going to toss a number cube that has the numbers 1, 2, 3, 4, 5, and 6 on it. How can you describe the probability of tossing a 3 or a 4? How can you describe the probability of tossing a number greater than 6? How can you describe the probability of tossing a number less than 7?

2. Five friends each ordered 1 scoop of ice cream and 1 topping. The store sells vanilla and chocolate ice cream, and hot fudge and butterscotch toppings. Can each of the friends get a different sundae? Explain.

3. The attendance each day for the Main Street library is listed in the table.

Day	Visitors
Monday	80
Tuesday	20
Wednesday	40
Thursday	60
Friday	30

a. Display the number of visitors in an appropriate graph.
b. How many people visited the library Monday through Friday?

4. James, Kyle, and Lizzie are running for the student council. The student with the most votes will be president. The student with the next highest number of votes will be vice-president, and the other student will be secretary. In how many different ways can the places be filled? What are they?

5. Cramer Elementary School is putting on a Thanksgiving show. The school is offering a prize to the student who sells the most tickets to the show. The table shows the number of tickets sold by the top 4 students.

Name	Adult Tickets Sold	Child Tickets Sold
John	12	8
Martina	13	8
Joanie	7	24
Harold	12	16

a. Who sold the most tickets?
b. How many tickets were sold in all?
c. Of which type of ticket did they sell more? How many more?

6. Rick has been given the following data set.

6, 7, 9, 5, 4, 6, 3

Which of the following does 6 represent: range, median, mode?

7. Nikki has number tiles with the digits 3, 5, and 8. If she makes 3-digit numbers from those tiles, how many different numbers can she make?

10. Test #1

Answer the following questions to the best of your ability. Remember, even if you are unsure of how to solve the problem, you will always earn some credit if you begin the problem. Good luck!

1. Mary and Dan went on the bobsled, the haunted house, and the whip. What is the probability that they went on the whip first, the bobsled second, and the haunted house third?

2. Barbara is digging a flower garden in the shape of a rectangle. She wants the garden to be 10 feet long. She has 28 feet of fencing to enclose the garden. How wide will her garden be?

3. Juanita has 19¢ in her pocket. How many different combinations of coins could she have?

4. Meg has 16 fewer coins than Nancy. Nancy has 13 more coins than Laura. Laura has 10 more coins than Kris. Kris has 112 coins. How many coins does Meg have?

5. How many fewer students were born in January, February, and March, than in May and June?

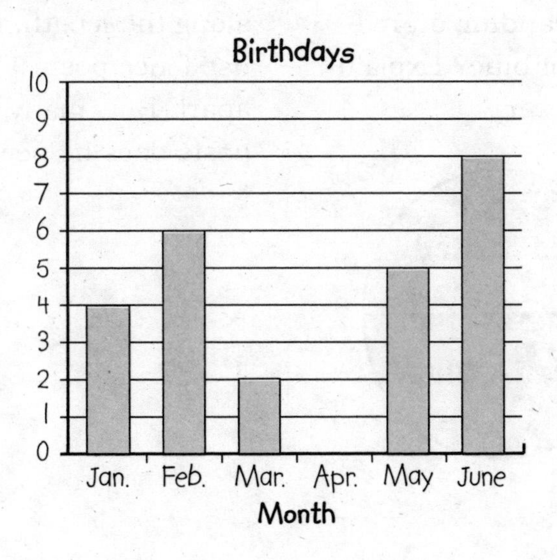

6. Lenny drew a square and a rectangle. The length of the square is less than the length of the shorter sides of the rectangle. If he combines the 2 figures, what polygon did he draw? Explain your answer.

10. Test #1

Answer the following questions to the best of your ability. Remember, even if you are unsure of how to solve the problem, you will always earn some credit if you begin the problem. Good luck!

7. Frankie is taking photographs in color and in black and white. She takes 3 rolls of color film with 24 photos in each roll. She takes 2 rolls of black-and-white film with 36 photos in each roll. Which type of film does she use to take more photos? How many more?

8. Dana has made this spinner. How can you describe the probability of the spinner landing on red instead of blue? Explain your answer.

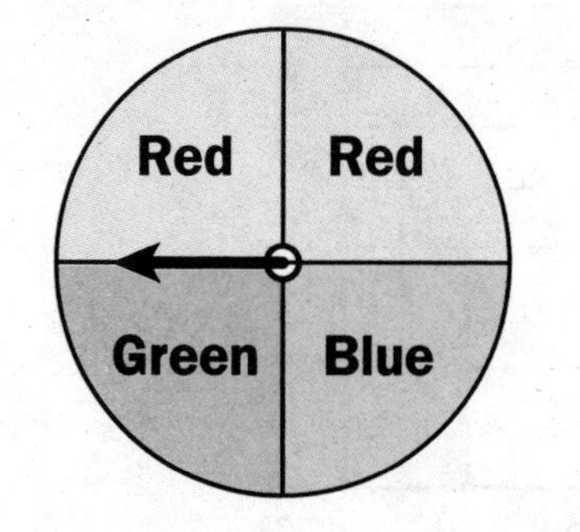

9. Martin's garden is a rectangle that is 15 feet long. He wants to put a fence along the length. He will use fence posts 1 yard apart. How many fence posts does he need?

10. Jerry is playing with a number machine. The table shows what has happened so far.

Input	Output
2	8
4	10
8	14
10	16
13	?

What is the missing number? Explain how you found your answer.

11. How can you find the lines of symmetry of a square? How many lines of symmetry does a square have?

12. The tallest player in NBA history stood 7 feet 7 inches. How tall is that player in inches?

13. The calendar for April is shown below.

APRIL

S	M	T	W	T	F	S
1	2	3	4	5	6	7
8	9	10	11	12	13	14
15	16	17	18	19	20	21
22	23	24	25	26	27	28
29	30					

Arlene left for a trip on April 12. She returned home 10 days later. What day of the week was this?

Answer the following questions to the best of your ability. Remember, even if you are unsure of how to solve the problem, you will always earn some credit if you begin the problem. Good luck!

14. Mr. Barker gave this number puzzle to his class. "I am thinking of 2 numbers. Their sum is 24 and their difference is 14. What are my 2 numbers?"

What 2 shapes form an angle?

16. Andy drew a rectangle that is 3 centimeters long and 1 centimeter wide. What does Sharon need to do to draw a rectangle that is similar but not congruent to Andy's?

17. Fernando is painting a design. The design consists of 2 triangles, 3 circles, and 4 squares. Then the pattern repeats over and over. What is the 25th shape of the design?

18. Dan has a tom-tom and a snare drum. He hits the tom-tom on every fourth beat. He hits the snare drum on every third beat. What is the first beat that Dan will hit both drums together?

19. Bev sells belts, wallets, key rings, and bags. Belts sell for $4. Wallets sell for $5. A key ring sells for $6 and a bag sells for $9. Ron bought 3 different items for exactly $15 from Bev. Which items did Ron buy?

20. Ralph runs 3 laps the first day, 7 laps the second day, and 11 laps the third day. If the pattern continues, how many laps will Ralph run on the seventh day?

10. Test #2

Answer the following questions to the best of your ability. Remember, even if you are unsure of how to solve the problem, you will always earn some credit if you begin the problem. Good luck!

1. Bonnie went on the Spaceship, the roller coaster, and the Alien Monster ride. What is the probability that she went on the Spaceship first, the Alien Monster ride second, and the roller coaster third?

2. Anita has a vegetable garden in the shape of an isosceles triangle. The 2 equal sides are each 15 feet long. The base is 10 feet long. She has 12 yards of fencing to enclose her garden. Can she do it? Why or why not?

3. Justin has 13¢ in his pocket. What coins could he have?

4. Scott is 3 years older than Brian. Brian is 4 years older than Jean. Jean is 2 years younger than Miranda. Miranda is 7 years old. How old is Scott?

5. The bar graph below shows the favorite flavor of ice cream as voted by Ms. Finnegan's class.

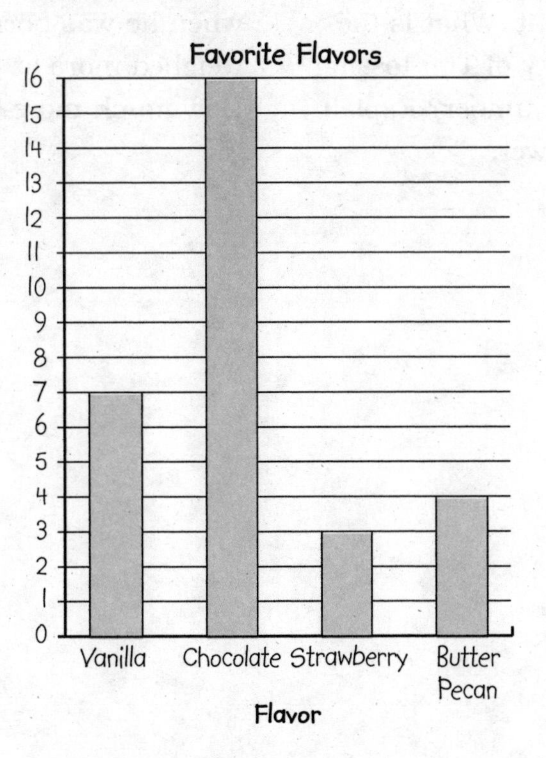

How many more students voted for chocolate than for all of the other flavors combined?

6. Which shape can also be classified as a rectangle? Explain your answer.

10. Test #2

Answer the following questions to the best of your ability. Remember, even if you are unsure of how to solve the problem, you will always earn some credit if you begin the problem. Good luck!

7. Martina works at the ballpark candy stand. She earns $8 an hour. Last Sunday she worked from 1:15 p.m. to 7:15 p.m. How much money did she earn?

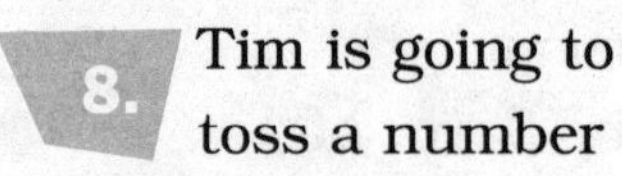

8. Tim is going to toss a number cube that has the numbers 1, 2, 3, 4, 5, and 6 on it. What is the probability of Tim tossing an even number? Explain your answer.

9. Louise weighed 7 pounds 8 ounces when she was born. Stan weighed 105 ounces when he was born. Who weighed more at birth? How much more?

10. Jeff is playing with a number machine. The table shows what has happened so far.

Input	Output
2	6
5	15
7	21
9	27
11	?

What is the missing number? Explain how you found your answer.

11. How can you find the lines of symmetry of an equilateral triangle? How many lines of symmetry does an equilateral triangle have?

12. Mr. Snyder is 77 inches tall. How tall is he in feet and inches?

13. Johanna is planning her birthday party. Today is February 7 and her birthday is in exactly 17 days. On what day of the week is her birthday?

FEBRUARY

S	M	T	W	T	F	S
		1	2	3	4	5
6	7	8	9	10	11	12
13	14	15	16	17	18	19
20	21	22	23	24	25	26
27	28					

10. Test #2

Answer the following questions to the best of your ability. Remember, even if you are unsure of how to solve the problem, you will always earn some credit if you begin the problem. Good luck!

14. Ms. Garrett gave her third graders a number puzzle. This is her puzzle. "I am thinking of 2 numbers. Their sum is 28 and their difference is 12. What are my numbers?"

15. Mrs. Brown gave her class this description of a figure. It has 3 sides. One angle is greater than a right angle, and 2 of the sides are equal. What figure did Mrs. Brown describe?

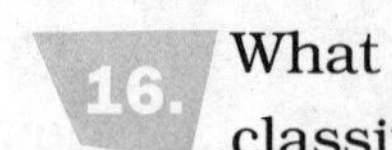

16. What classification of triangle is always similar to all others of the same type? Explain your answer.

17. Chen is painting a design of 3 triangles, 3 circles, and 2 squares. The pattern repeats over and over. What is the 21st figure?

18. Gladys changes the water in her guppy tank every second week. She changes the water in her beta tank every fifth week. What is the first week that she will change the water in both tanks?

19. At the Food Snack Shop a salad sells for $3. A turkey wrap sells for $5. A chicken basket sells for $7. A dessert sells for $4. Mike bought 3 different items and spent $14. What did he buy?

20. Beth is increasing the number of minutes she runs each day. On the first day, she ran for 15 minutes. On the second day, she ran for 25 minutes. On the third day, she ran for 35 minutes. If her pattern continues, how many minutes will Beth run on the seventh day?

11. Home-School Connection

Working on these questions at **home** with a **family member** is fun! Find a comfortable place to work and have all the tools you need. Discuss how you want to solve the open-ended math question. Then go for it! Don't forget to use your rubric!

11. Home-School Connection

Dear Family Member:

This year your child will be learning about **open-ended questions** in mathematics class. An open-ended question is a mathematics word problem that has one correct answer, but that can be solved in several different ways. Open-ended questions are extremely important on tests your child will take.

You can help your child practice solving these questions by working together on the take-home sheets in this chapter. Don't forget to use the rubric as a guide. Remember, when you work with your child, don't do the problem for her or him. Rather, encourage your child to ask questions that will lead him or her to the answer. And don't be surprised if your child arrives at the answer to the problem using a method different from the one you're thinking of.

It is important that your child make his or her thinking clear to the reader. After your child has solved the problem, (or when you child has gone as far as he or she can) help your child write a clear explanation of what he or she did to solve it, and why he or she decided to do it that way. This will help your child clarify his or her own thoughts.

The problems on the following pages are based on the areas of mathematics considered important in solving open-ended problems. These are:

- **Number and Operations**
- **Algebra**
- **Geometry**
- **Measurement**
- **Data Analysis and Probability**

Enjoy!

Number and Operations

Problem

Students are planting a garden to have some flowers in the schoolyard. They planted 3 rows of rose bushes, with 7 bushes in each row. Then they planted 4 rows of lilac bushes with 5 bushes in each row. How many bushes did they plant all together?

Algebra

Problem

George is playing with a number machine. The table shows what has happened so far.

Input	Output
3	6
5	10
8	16
11	22
15	?

What is the missing number? Explain how you found your answer.

Geometry

Problem

Ms. Roberts described to the students of her class this mystery figure.

"The figure I am thinking of has 3 dimensions with 6 faces. All of faces are squares. What figure am I thinking of?"

Explain your answer.

Measurement

Problem

For a picnic, Mr. Rollins brought 12 quarts of fruit juice. How much fruit juice did he bring, in gallons? Explain your answer.

Data Analysis and Graphs

Problem

The line graph shows the temperatures each hour from 9 a.m. until 12 noon. How much warmer was it at 11 a.m. than at 9 a.m.?

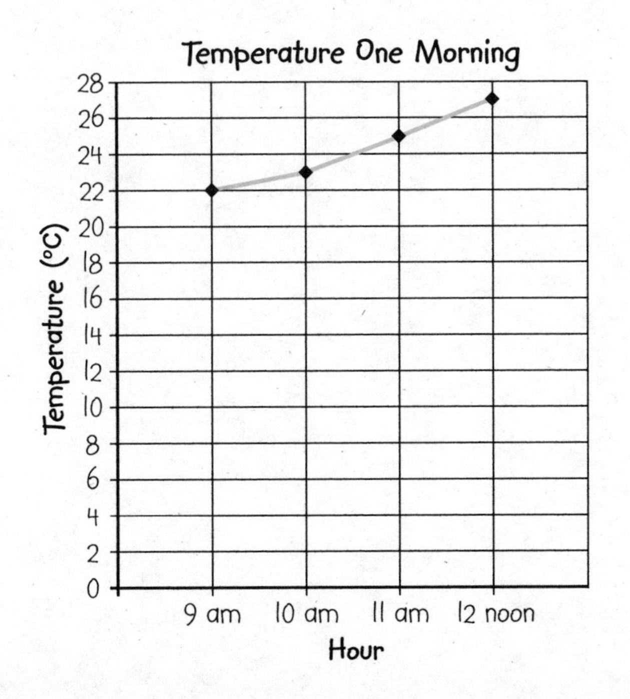

Glossary

A

After At a later time.

Ahead In front of.

Algebra The branch of mathematics that uses letters and symbols to write expressions about number relationships.

All together With all included.

Apart A distance between 2 objects.

Area The number of square units needed to cover a region.

Arrived Having reached a destination.

Glossary

B

Bar graph A graph that shows data by using bars of different lengths.

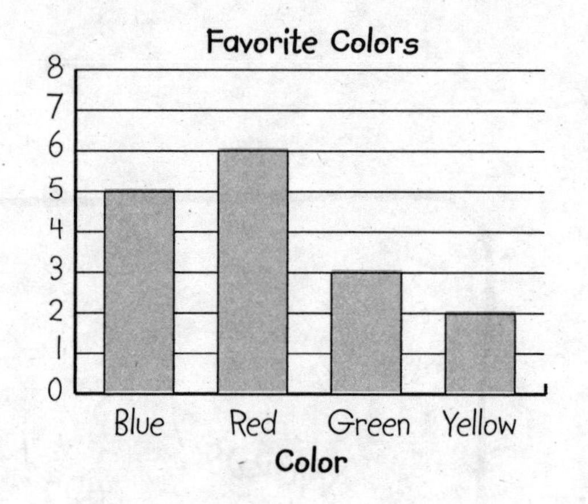

Behind Toward the rear.

C

Cent (¢) A unit of money. One hundred cents equals one dollar.

Centimeter A metric unit of length equal to $\frac{1}{100}$ of a meter.
—— This line segment is 1 centimeter long.

Chance The possible outcome of an event.

Circle A closed 2-dimensional figure having all points the same distance from a given point.

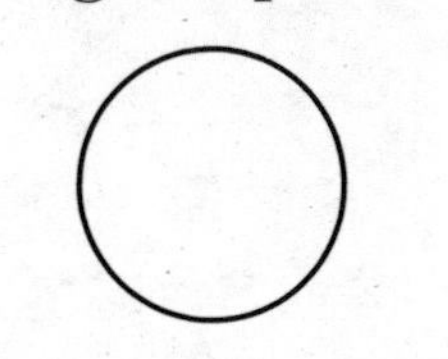

Combinations Groups.

Combined Joined.

Cone A pointed 3-dimensional figure with a circular base.

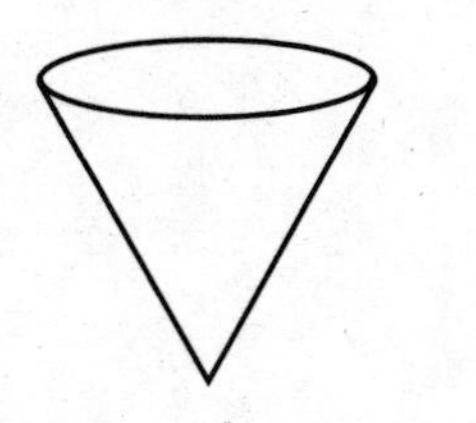

Congruent A figure having the same size and shape as another figure.

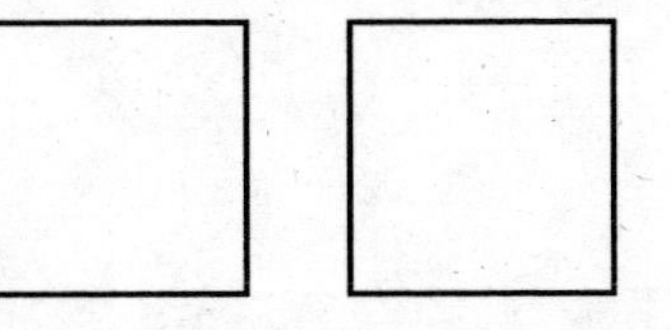

These squares are congruent.

Glossary

Coordinate grid A grid used to show location.

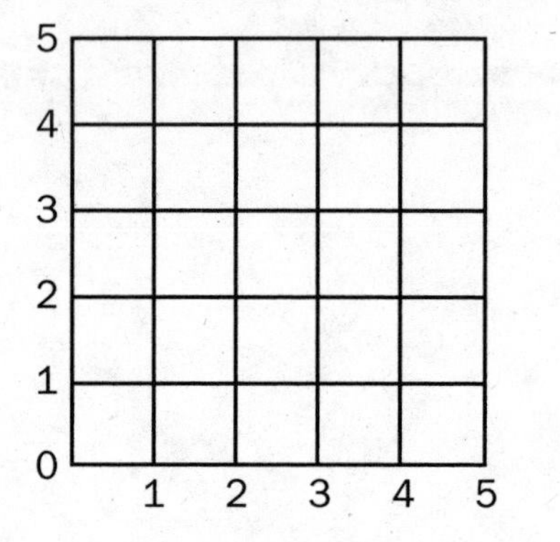

Corner One of the vertices of a rectangle.

Cost To have a certain price.

Crosses Meets.

Cube A 3-dimensional figure with six congruent square faces.

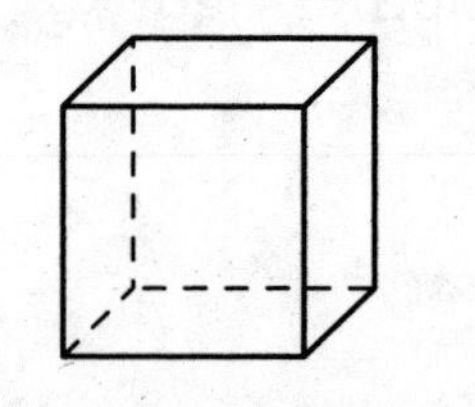

Cup A customary unit of capacity.
2 cups are equal to a pint.

Cylinder A 3-dimensional figure with 2 congruent faces that are circular.

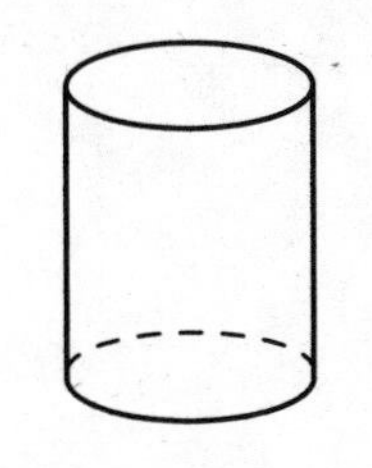

D

Data Information.

Date An exact day on a calendar.

Day A unit of time lasting 24 hours.

Degree Unit for measuring temperature.
Celcius (°C) Metric Unit for measuring temperature. Freezing point 0°C, Boiling Point 100°C
Fahrenheit (°F) Customary unit for measuring temperature. Freezing point 32°F, Boiling Point 212°F

Difference The answer in a subtraction problem.

Different Not the same.

Dime A coin worth 10¢. It is worth $\frac{1}{10}$ of a dollar.

Dollar ($) A bill or coin worth 100¢. It is referred to as $1 or $1.00.

E

Each Every one of a group considered individually.

Enough An adequate quantity.

Equally The same number in each group, equal groups.

Equation A number sentence that says 2 values are equal.
Examples: $2x = 8$, $2 + 2 = 4$.

Equilateral triangle A triangle with 3 congruent sides and angles.

F

Factor One of 2 numbers that are multiplied to give a product.
Example: In $5 \times 6 = 30$, 5 and 6 are factors.

Finished The end.

Foot A customary unit of length equal to 12 inches.

Form To create.

Fraction A number that names part of a whole or group. Example: $\frac{2}{3}$.

G

Geometric Having to do with geometry.

Geometry The study of figures.

Graph A picture that displays the data.

Greater An amount larger than another one.
Example: 12 is greater than 7.

Guess and Test A problem-solving strategy in which you make a guess, test your guess, and then change your guess until finding the correct answer.

H

Hour A unit of time that contains 60 minutes.

Glossary

I

In all With all included.

Inch A customary unit of length equal to $\frac{1}{12}$ of a foot.
__________ This line segment is 1 inch long.

Isosceles Triangle A triangle with 2 equal sides.

K

Kilometer A metric unit of length equal to 1,000 meters.

L

Later After.

Left Departed.

Length The measurement of distance between 2 endpoints.

Less Fewer.

Line of symmetry A line on which a figure can be folded so that both sides match.

Line segment A straight path that has 2 endpoints.

Longer To take a greater amount of time.

M

Many A large number.

Measure The dimensions of a figure.
The length of a rectangle is its measure.

Meter A metric unit of length equal to 100 centimeters.

Minute A unit of time equal to 60 seconds.

More Greater in number or amount.

Most More than $\frac{1}{2}$.

Much An amount.

N

Next Immediately following.

Nickel A coin worth 5¢.

Now At this time.

Number sentence An equation.

O

Open-ended problem A math problem that has a correct answer, that you can arrive at in more than one way.

Order An arrangement.

Example: From least to greatest.

Ordered pair A pair of numbers that gives the location on a coordinate grid.

Ounce A customary unit of weight equal to $\frac{1}{16}$ of a pound.

P

Parallel lines Lines that never meet and remain the same distance apart.

Pattern A series of numbers or figures that follows a rule.

Perimeter The distance around a figure. The perimeter is found by adding the lengths of all of a polygon's sides.

Perpendicular lines Lines that meet at a right angle.

Pint A customary unit of capacity equal to 2 cups.

Plotted Drew.

Polygon A closed geometric figure with all sides straight lines.

Pound A customary unit of weight equal to 16 ounces.

Glossary

Probability The chance of an event occurring.

Product The answer in a multiplication problem.

Q

Quart A customary unit of capacity equal to 2 pints.

Quarter A coin worth 25¢. It is equal to $\frac{1}{4}$ of a dollar.

R

Rectangle A polygon with 4 sides, 4 right angles, and 2 pairs of opposite congruent sides.

Rectangular Having the shape of a rectangle.

Repeat To do again.

Rest An amount that remains.

Right angle — An angle that forms a square corner.

Roll — To turn over and over.

Rose — Increased.

Rule — A statement that explains a pattern.

S

Shape — Form, figure.

Shared — Divided equally.

Side — One of the line segments in a polygon.

Glossary

Similar Figures that have the same shape, but may have different sizes.

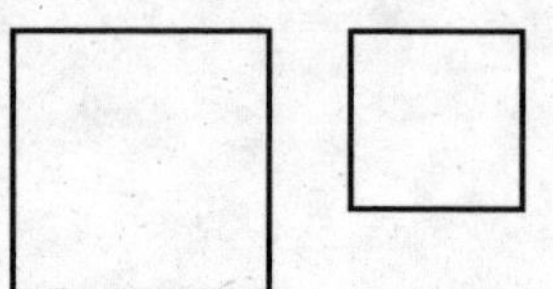

These squares are similar.

Single One.

Specifically As exact as possible.

Sphere A 3-dimensional figure that has the shape of a ball.

Square A rectangle with 4 equal sides.

Started The beginning.

Sum — The answer in an addition problem.

Symbol — Something that stands for something else. For example, the symbol + means addition.

T

Table — A display used to show data without a picture.

Temperature — A measurement that tells how hot or cold something is.

3-dimensional figure — A figure that has length, width, and height.

Times — Multiplied by.

Triangle — A polygon with three sides and three angles.

Triangular — Having the shape of a triangle.

Twice — Two times.

U

Unit — A thing that is part of a group.

V

Variable A value that can change or needs to be found.

Vertex A point where 2 rays meet; a point where 2 sides of a polygon meet.

W

Week A unit of time equal to 7 days.

Weight A measurement that tells how heavy an object is.

Width The length of the shorter sides of a polygon such as a rectangle.

Y

Year A unit of time equal to 365 days.

Notes

Notes